Translations of Mathematical Monographs

Volume 46

Projection-Iterative Methods for Solution of Operator Equations

by

N. S. Kurpel′

American Mathematical Society
Providence, Rhode Island
1976

ПРОЕКЦИОННО-ИТЕРАТИВНЫЕ МЕТОДЫ
РЕШЕНИЯ ОПЕРАТОРНЫХ УРАВНЕНИЙ

Н. С. КУРПЕЛЬ

Издательство "Наукова Думка"
Киев–1968

Translated from the Russian by
Israel Program for Scientific Translations

Translation edited by R. G. Douglas

AMS (MOS) subject classifications (1970). Primary 47A50, 47H15;
Secondary 65J05, 65H10, 65R05, 34C15, 73K15.

Abstract. In this monograph general approximation methods for solving linear and nonlinear operator equations are studied, combining the ideas of projection as well as iterative methods and effecting further generalizations of the method of averaging functional corrections. Part of the book is devoted to the application of these methods to various special classes of operator equations.

Library of Congress Cataloging in Publication Data

CIP

Kurpel', Nikolai Stepanovich.
Projection-iterative methods for solution of operator equations.

(Translations of mathematical monographs ; v. 46)
Translation of Proektŝionno-iterativnye metody resheniîa operatornykh uravneniĭ.
Bibliography: p.
1. Operator equations. 2. Iterative methods (Mathematics) I. Title. II. Series.
QA329.K8713 515'.72 76-17114
ISBN 0-8218-1596-2

TABLE OF CONTENTS

Preface

The general approximate methods considered in this monograph for the solution of operator equations, called iterative projection methods, constitute a further generalization of an effective method of Sokolov for the approximate solution of various types of equations, now known in the literature as the method of averaged functional corrections. This method has already found extensive application in the solution of diverse problems of mathematics and mechanics. For a theoretical justification and practical applications of the method and some of its modifications and generalizations, the reader may consult the detailed monographs of Sokolov [197] and Lučka [120], apart from numerous papers.

This book is based on results of the author's research. Other results are cited, for the most part, only when they have a direct bearing on the material. Part of the book is devoted to a detailed exposition of results published previously in abbreviated form; the other part contains results published here for the first time.

The monograph comprises an introduction and three chapters.

The introduction is a brief survey of the most familiar iterative and projection methods for sulution of operator equations. It presents the gist of the method of averaged functional corrections and a brief survey of papers investigating these methods.

In Chapter I we set up a general scheme of iterative projection methods for operator equations in generalized metric and lattice-normed spaces, in which the distances and norms are elements of certain semiordered spaces. We study convergence criteria and establish error estimates for sequences of approximations defined by iterative projection methods, both for the general algorithm and for some of its specializations.

Chapter II is devoted to the investigation of various specialized iterative projection methods in special metric and normed spaces, sufficient conditions for convergence and error estimates which do not follow directly from the general theory of Chapter I.

Finally, in Chapter III the results of the previous chapters are applied to various special types of operator equation—systems of algebraic and transcendental equations, integral and differential equations, and systems of such equations. Simple examples are presented in an illustrative capacity.

For an understanding of the main text, the reader should be familiar with the basics of functional analysis in normed and semiordered spaces. However, some of this information may be drawn from special introductory sections devoted to auxiliary material.

The bibliography provided at the end of the book is fairly complete with regard to literature on the method of averaged functional corrections, its modifications, generalizations and applications, as well as directly related material.

The author is deeply indebted to Ju. D. Sokolov, Corresponding Member of the Academy of Sciences of the Ukrainian Soviet Socialist Republic, for his constant interest in this work and his invaluable remarks and advice. Thanks are also due to A. Ju. Lučka, who read the manuscript and offered many valuable comments.

INTRODUCTION

At the present time, the solution of many problems of natural science and technology would be unthinkable without the extensive use of approximate and numerical methods. Of the great wealth of such methods, the best known are difference, iterative and projection methods. We shall not spend time characterizing difference methods, which are now a standard tool both in proving existence theorems for solutions of various classes of differential equations and in the actual determination of solutions; the reader may consult [41], [52], [107] and the detailed survey [60], all of which contain, in particular, extensive bibliographies.

The idea of the method of successive approximations or the iterative method originated in the eighteenth century. It was used by Liouville in his studies of linear differential equations as far back as 1836, and then by C. Neumann in potential theory [144]. Subsequently, the method was utilized with increasing frequency both to prove existence theorems for various types of equations and to determine approximate solutions.

The functional-analytic approach to approximate methods for solution of equations provided a general framework for consideration of the iterative method. Since many important classes of equations which may be treated by successive approximations are special cases of the general operator equation

$$x = Tx, \tag{1}$$

where T is an operator defined in some function space E, it is quite natural to investigate the iterative method as applied directly to equation (1). As applied to this equation, the usual method of successive approximations starts from some initial approximation x_0, which may be selected arbitrarily in E, and then defines successive approximations $x_1, x_2, \ldots$ to the solution by the recurrence relation

$$x_n = Tx_{n-1} \quad (n = 1, 2, \ldots). \tag{2}$$

Let E be a complete linear metric space and T a contractive operator in the space, i.e., for any $u, v \in E$

$$d(Tu, Tv) \leqslant \alpha d(u, v), \tag{3}$$

where $\alpha < 1$ ($d(x, z)$ is the distance between x and z).

Then equation (1) has a unique solution x^* in E, which is the limit of the sequence $\{x_n\}$ whose terms x_n are defined by (2). The error of the nth approximation x_n satisfies the following estimate:

$$d(x^*, x_n) \leqslant \frac{\alpha^{n-p}}{1-\alpha} d(x_{p+1}, x_p) \quad (0 \leqslant p \leqslant n-1). \tag{4}$$

This theorem was established (in a slightly different form) by Banach in 1922 [8] and subsequently became known as the Banach contraction mapping principle.

Thanks to the efforts of many mathematicians, the method of successive approximations is now the basis of many proofs of existence and uniqueness theorems for solutions of various problems of mathematical physics.

In cases where the conditions for applicability of the contraction mapping principle are not fulfilled (in particular, condition (3)), one often employs the Schauder fixed-point principle [167]: If T is a continuous mapping of a convex set R in a Banach space into a compact subset, then T has at least one fixed point in R, i.e., equation (1) has at least one solution in R.

Applications of fixed-point principles to various types of nonlinear integral equations have been considered by Nemyckiĭ [142], [143], Nazarov [141], Smirnov [180] and many others.

A highly significant development of the method of successive approximations was achieved by Kantorovič, thanks to his theory of semiordered spaces (see [71] and [229]). This theory made it possible to develop and justify certain other methods as well.

Using the theory of semiordered spaces, Baluev [7] constructed an abstract framework for the method of Čaplygin [28] for the solution of ordinary differential equations. Similar results were then obtained by Slugin [179].

Many authors have applied the method of successive approximations to approximate solution of various special classes of equations in function and vector spaces. Of the numerous publications on this subject, we might mention Ważewski [232]–[234], Collatz [36]–[39], Weissinger [235], and Schröder [172]–[174].

Further generalizations of fixed-point principles may be found in Tihonov [202], Krasnosel'skiĭ [77], [78], Vaĭnberg [223], Kačurovskiĭ [64], Browder [20], Albrecht and Karrer [1], Edelstein [48], Schaefer [166], and others.

Apart from stationary iterative methods, characterized by the fact that the operator T is the same at each step, there has been some interest in the so-called nonstationary iterative methods, in which the operator T may be different at each step. In abstract form, the method may be expressed as

$$x_n = T_n x_{n-1} \quad (n = 1, 2, \ldots), \tag{5}$$

where $\{T_n\}$ is a sequence of operators converging to an operator T as $n \to \infty$. Ehrmann [50] has justified the algorithm (5) for the case that equation (1) is considered in a generalized metric space with the distance an element of a semi-ordered space.

One of the most widespread iterative methods other than the ordinary method of successive approximations is Kantorovič's approximate method for operator equations, which generalizes Newton's method of tangents and is known in the literature as the Newton-Kantorovič method. According to this method, the successive approximations x_n to the solution of an operator equation

$$F(x) = 0 \tag{6}$$

are defined by

$$x_n = x_{n-1} - [F'(x_{n-1})]^{-1} F(x_{n-1}) \quad (n = 1, 2, \ldots) \tag{7}$$

(on the assumption that at each point x_{n-1} the Fréchet derivative $F'(x)$ of F and its inverse $[F'(x_{n-1})]^{-1}$ exist).

Kantorovič also proposed what he called a modified Newton method, defining the successive approximations by

$$x_n = x_{n-1} - [F'(x_0)]^{-1} F(x_{n-1}) \quad (n = 1, 2, \ldots). \tag{8}$$

In contradistinction to method (7), which is characterized by quadratic convergence, this method guarantees convergence at the rate of a geometric progression.

Apart from Kantorovič himself [67], [68], many other authors have studied methods (7) and (8).

Both the basic and modified Newton-Kantorovič methods are equivalent to the ordinary method of successive approximations, applied to the respective equations

$$x = x - [F'(x)]^{-1} F(x) \qquad \text{and} \qquad x = x - [F'(x_0)]^{-1} F(x).$$

Under the influence of Kantorovič's ideas, many papers then appeared investigating iterative methods with higher orders of convergence.

Of the other iterative methods for solution of operator equations, we mention the method of steepest descent [67], [69], [224], the iterative process with minimal residuals [79], and the iterative method with minimal errors [55].

A broad range of approximate methods may be subsumed under the heading of projection methods. Among these we find, in particular, the methods of Ritz [162], [163], Galerkin [59], least squares, moments, and some of their modifications and generalizations.

The generalized version of the Ritz method is applied to the nonlinear operator equation (6), where the operator F is the gradient of some functional Φ with a linear domain of definition $D(\Phi)$ dense in some separable space E.

The main idea of the method is to reduce the solution of equation (6) to the minimization of Φ on a set of elements of $D(\Phi)$; the minimizing element is sought in the form

$$x_n = \sum_{i=1}^{n} a_i \varphi_i, \tag{9}$$

where $\varphi_1, \varphi_2, \ldots$ are elements of $D(\Phi)$, linearly independent for each n and forming a complete system in E; the constants a_i are determined by the condition that Φ be a minimum on the set of elements (9). If Φ is continuously differentiable on any finite-dimensional hyperplane in its domain of definition, the coefficients a_i are solutions of the system of equations

$$\frac{\partial \Phi\left(\sum_{i=1}^{n} a_i \varphi_i\right)}{\partial a_j} = 0 \quad (j = 1, 2, \ldots, n). \tag{10}$$

As applied to various problems of mathematical physics, the Ritz method was investigated by Krylov and Bogoljubov [87] – [89], Kantorovič [67] and others. The general case of nonlinear operator equations has been considered in [47], [128] and others.

For the solution of many problems of mathematical physics and mechanics, the much more general method of Galerkin is more widely used than the Ritz method. Applied, say, to the operator equation (6) in a Hilbert space H, this method assumes as before that the approximate solution is given by (9), but the constants a_i are determined by the condition that the residual $F(x_n)$ be orthogonal to the elements $\varphi_1, \ldots, \varphi_n$: that is, by the system of equations

$$\left(F\left(\sum_{i=1}^{n} a_i \varphi_i\right), \varphi_j\right) = 0 \quad (j = 1, 2, \ldots, n). \tag{11}$$

For many years, the Galerkin method was used to solve various applied problems without adequate mathematical justification. For a certain special problem, Petrov [146] established the convergence of the Galerkin successive approximations. General results on convergence of the Galerkin method for boundary-value problems were obtained by Keldyš [72]. Subsequently, functional analysis came into use as a tool for studying the Galerkin method and other projection methods. In this connection, the fundamental results on convergence of projection methods are due to Kantorovič [67], Mihlin [127], [128], Pol'skiĭ [156] – [159], and Krasnosel'skii [76], [78].

The method of moments is more general; it differs from the Galerkin method in that the constants a_i are determined by the condition that the residuals $F(x_n)$ be orthogonal to the elements of a certain other system of linearly independent elements $\{\psi_i\}$: that is, by the system

$$\left(F\left(\sum_{i=1}^{n} a_i\varphi_i\right),\ \psi_j\right) = 0 \quad (j = 1, 2, \ldots, n). \tag{12}$$

A broad range of problems may be dealt with through the so-called method of least squares. Here one looks for the approximate solution to equation (6) in the form (9), with the constants a_i minimizing the norm of the residual, $\|F(x_n)\|$. Moreover, if $\|F(x_n)\|$ is a continuously differentiable function of the parameters a_i, the latter are determined from the system

$$\frac{\partial \left\| F\left(\sum_{i=1}^{n} a_i\varphi_i\right)\right\|}{\partial a_j} = 0 \quad (j = 1, 2, \ldots, n). \tag{13}$$

The method of least squares was applied to differential and integral equations in [87]–[89] and other publications. The methods of moments and least squares were investigated by Kravčuk [82]. Mihlin [126], [127] gave a theoretical justification of the method of least squares for linear operator equations, and also applied it to various applied problems. Of the more recent papers on the least squares method for nonlinear operator equations, we mention Langenbah [108].

Note that if equation (6) is a differential equation and we consider the corresponding boundary-value problem, the functions φ_i are usually chosen so as to satisfy the appropriate boundary conditions.

We consider a projection method which is specifically applicable to the linear operator equation in a separable Hilbert space H:

$$Ax - f = 0 \quad (f \in H). \tag{14}$$

One assume that the approximate solution will have the form

$$x_n = \sum_{i=1}^{n} a_i A^*\varphi_i, \tag{15}$$

where $\{A^*\varphi_i\}$ is a complete linearly independent system in H, A^* is the adjoint of A, and the constants a_i are determined so as to minimize the norm $\|x^* - x_n\|$ (where x^* is the exact solution of equation (14)). The values of a_i minimizing $\|x^* - x_n\|$ are determined from the system

$$\sum_{i=1}^{n} a_i (A^*\varphi_i, A^*\varphi_j) - (f, \varphi_j) = 0 \quad (j = 1, 2, \ldots, n). \tag{16}$$

This method is described by Fridman [55]. Incidentally, it is a generalization of the well-known method of Enskog for solution of integral equations [215].

Thanks to functional analysis, projection methods may be considered from a unified standpoint and incorporated in a general theory. Indeed, each of the projection methods listed above may be interpreted as a certain way of replacing the original equation in the space E by a new equation, "close" in some sense to the former, but considered in a certain finite-dimensional subspace of E. Since the new equation is "close" to the original one, it is natural to expect its solution to approximate that of the original equation.

In this general setting, the question of convergence conditions for projection methods in certain classes of linear operator equations was studied by Pol′skiĭ [156]–[159], Mihlin [126] and others.

Krasnosel′skiĭ in [76] and [78] uses topological methods to investigate the convergence of projection methods for equations of type (1), where T is a compact nonlinear operator defined in some Banach space. The approximate solution of equation (1) is defined as a solution of an approximating equation

$$x_n = P_n T x_n, \tag{17}$$

where P_n is the projection of the space E onto the subspace E_n spanned by the first n elements of some basis $\{\varphi_i\}$. Krasnosel′skiĭ proved that if equation (1) has an isolated solution x^* of nonzero index, then the approximate solutions x_n exist for sufficiently large n and converge as $n \longrightarrow \infty$ to the solution x^* of equation (1). If the operator T has a Fréchet derivative at x^* for which 1 is not an eigenvalue, then the rate of convergence of the sequence $\{x_n\}$ is given by

$$\| x^* - x_n \| \leqslant (1 + \varepsilon_n) \| (P_n - I) x^* \|, \tag{18}$$

where $\epsilon_n \longrightarrow 0$ as $n \longrightarrow \infty$, I is the identity operator.

The projection method (17) may be regarded as a special case of a more general procedure, according to which equation (1) is replaced by an approximating equation

$$x = Sx, \tag{19}$$

where S is an operator which is in some sense close to T, and the solution of this equation is taken as the approximate solution of equation (1). This general method is due to Kantorovič [67], [69]. Projection methods have recently been investigated by Petryshyn [147] via the theory of K-positive-definite operators.

A few words about a problem of great practical importance, namely, the estimation of the error incurred by replacing equation (1) by equation (19). The general problem has been considered by Ehrmann [50] and Schmidt [169]–[171]. In particular, Schmidt [169] proves the following theorem.

Suppose that the operator T, defined in a space E normed by elements of an archimedean set N, maps some complete subset D of E into itself, and for any $y, z \in D$

$$\| Ty - Tz \| \leqslant L \| y - z \|, \tag{20}$$

where L is a positive linear operator defined in N and the series $\Sigma_{i=0}^{\infty} L^i \rho$ converges uniformly for any $\rho \in N$. Then T has a unique fixed point $x^* \in D$ and we have the error estimate

$$\| x^* - v \| \leqslant (I - L)^{-1} \| (T - S) v \|, \tag{21}$$

where v is a solution of equation (19).

An analogous assertion under other assumptions may be found in [50]. For linear integral equations, an effective error estimate was found previously by Kantorovič [70], and for nonlinear integral equations by Mysovskih [139].

Both iterative and projection methods have their specific advantages and disadvantages. The considerable interest evinced lately in iterative methods is due to the fact that, by virtue of their highly simple computational scheme, they are more easily handled on modern computers than other methods. However, their usefulness is limited to a relatively narrow range of problems, since the approximations do not always converge to a solution. Even when the process of successive approximations does converge to a solution, its application may be inefficient because of the overly large amount of computational labor required to achieve sufficient accuracy.

Projection methods enjoy a wider field of application. However, the determination of sufficiently accurate approximations by projection methods is frequently bound up with the need to solve systems of equations of high orders, a very difficult task in itself, especially in regard to nonlinear equations. In addition, the approximation obtained for $n = k$ is usually not used to find higher-order approximations ($n > k$). It is also extremely difficult to decide what values of n will yield the required accuracy, and to select the sequence of elements φ_i.

A natural development, generalizing and perfecting both iterative and projection methods, would be to devise new methods combining the ideas of both. Such methods are known as iterative projection methods.

One such method is "averaging of functional corrections," proposed in 1952 by Sokolov for the approximate solution of integral and differential equations. A detailed account of the principal version of the method may be found in Sokolov's papers [183]–[196] and his monograph [197].

We shall illustrate Sokolov's method for the case of a nonlinear integral equation

$$y(x) = \varphi(x) + \int_a^b K(x, \xi) f[x, \xi, y(\xi)]\, d\xi. \tag{22}$$

We define successive approximations $y_n(x)$ to the solution of equation (22) by

$$y_n(x) = \varphi(x) + \int_a^b K(x, \xi) f[x, \xi, y_{n-1}(\xi) + \alpha_n] \, d\xi \quad (n = 1, 2, \ldots), \tag{23}$$

where

$$\alpha_n = \frac{1}{b-a} \int_a^b \delta_n(x) \, dx, \; \delta_n(x) = y_n(x) - y_{n-1}(x), \; y_0(x) = 0. \tag{24}$$

To find the corrections α_n at each step, we have an algebraic or transcendental equation

$$(b-a)\,\alpha_n = \int_a^b \left\{ \varphi(x) + \int_a^b K(x, \xi) f[x, \xi, y_{n-1} + \alpha_n] \, d\xi - y_{n-1}(x) \right\} dx. \tag{25}$$

Subsequently, the method of averaged functional corrections was used by Sokolov to solve linear and nonlinear integral equations with variable [191], [192] and mixed [193] limits of integration, and also systems of linear [194] and nonlinear [195] integral equations.

Sokolov established several criteria for convergence of his method, and error estimates. In [192], for example, he derived a sufficient condition for convergence of the method (23)–(24) for equation (22), similar to the well-known convergence criterion for the standard method of successive approximations. This criterion was generalized to systems of integral equations in [195]. Less restrictive conditions and more accurate error estimates may be found in [196].

Sokolov's main results in regard to the averaging of functional corrections are summarized in his monograph [197].

Extending the idea of the method of averaged functional corrections, Černyšenko [29] constructed successive approximations to the solution of equation (1) in a complete normed space, by the formulas

$$x_n = T(x_{n-1} + \alpha_n), \tag{26}$$

$$\alpha_n = S\delta_n, \; \delta_n = x_n - x_{n-1} \quad (n = 1, 2, \ldots), \tag{27}$$

where S is some linear averaging operator, and proved the convergence of the process under the highly restrictive condition

$$q_T(1 + 2\|S\|) < 1, \tag{28}$$

where q_T is a Lipschitz constant for T and $\|S\|$ the norm of S. Treating linear operator equations in a Hilbert space [32], Černyšenko then defined the averaging operator S as orthogonal projection of the original space onto the subspace spanned by an orthonormal system of elements $\{\varphi_i\}$ $(i = 1, \ldots, k)$. Averaging of functional corrections was subsequently applied to solution of the Cauchy problem for ordinary differential equations [33], determination of eigenvalues of linear operators [30], and solution of a certain applied problem [34].

The method was applied to the Cauchy problem and boundary-value problems for linear integrodifferential equations by Krivošein [83]–[85].

In a series of papers [112]–[123], Lučka investigated Sokolov's method and some of its generalizations to linear operator equations in Banach and Hilbert spaces, and also to various special classes of equations. Among his most important results in this direction are necessary and sufficient conditions and also certain effective criteria for convergence of the method, and more accurate error estimates. Lučka has shown, in particular, that in many important cases Sokolov's method is more widely applicable than the usual method of successive approximations and converges much more rapidly.

A number of papers by Mosolov [131]–[137] present applications of Sokolov's method and its generalizations to some special classes of operator equations. In particular, he considers nonlinear integrodifferential equations and systems of such equations [131], [137], and loaded integral and integrodifferential equations [134], and also generalizes a modification of Sokolov's method [132], [133] first proposed by Molokovič in [129] and [130].

The present author's papers [90]–[103] are devoted to laying a rigorous basis for Sokolov's method as applied to nonlinear operator equations. We have established several convergence criteria and corresponding error estimates, and also constructed and investigated some general algorithms which generalize the method of averaged functional corrections.

For linear integral equations of Volterra and mixed type, some new variants of Sokolov's method have been proposed and studied by Tivončuk [203]–[212].

Averaging of functional corrections was applied to some types of integral equations with retarded argument by Barataliev and Krivošein (see [9]–[13] and [86]), and also Gal′ (see [56]–[58]).

Among other publications devoted to the Sokolov method, we mention a paper of Sabirov and Esajan [164] using the theory of semiordered spaces, Stonickiĭ [198], [199] who solves a countable system of linear integral equations, and Sirenko [178] and Tukalevskaja [216], who consider numerical implementation of the method.

Some problems of mechanics are treated by the method of averaged functional corrections in Sokolov [183]–[186], [188] and also in Švec et al. [200].

Closely related to the averaging of functional corrections as regards its underlying ideas is Vorob′ev's method [228] for accelerating the convergence of an iterative process.

The totality of approximate methods for solving various classes of operator equations is subsumed under the scheme of a general method of successive approximations, according to which the approximations x_n to the solution of a general operator equation

$$\Phi(x, x) = 0 \tag{29}$$

are defined as solutions of operator equations

$$\Phi_n(x_n, x_{n-1}) = 0, \tag{30}$$

where $\{\Phi_n\}$ is a sequence of operators which converge in a suitable sense to Φ as $n \longrightarrow \infty$. Nevertheless, the fact that the general scheme (30) is open to investigation does not eliminate the need to study various special cases, since this general framework cannot possibly take into account the specific properties of concrete algorithms and operators.

The special case of (30) in which $\Phi(x, y) = x - F(x, y)$ and $\Phi_n(x, y) = \Phi(x, y)$, in the general setting of an abstract space metrizable by the elements of some partially ordered set, has been considered by Ważewski [234] and also by Kwapisz [104], [105].

We cite here one general iterative method, approximating the solution of equation (1) by elements

$$x_n = T_n x_n \quad (n = 1, 2, \ldots). \tag{31}$$

The operators T_n, which may also depend on $x_0, x_1, \ldots, x_{n-1}$, are usually selected so that they are simpler than T. For equations in a Banach space, the algorithm (31) has been studied by Warga [231], and for equations in a space normed by elements of an archimedean lineal by Schmidt [168]. Some variants of the general iterative method have been considered by Bel′tjukov [14], Tukalevskaja [217]–[221], Dīdik [44], Kalaĭda and Sereda [66], Ehrmann [49], Frey [53], [54], Kowalski [75], Olech [145], Picone [150], [151], Todorow [214], Vo-Khac Khoan [227], Zuber [236]–[238], and others.

CHAPTER I

GENERAL THEORY OF ITERATIVE PROJECTION METHODS FOR OPERATOR EQUATIONS IN ABSTRACT METRIC AND NORMED SPACES

§1. Auxiliary material from the theory of semiordered spaces

1. *Partially ordered sets.* A set is said to be partially ordered if there is a relation $<$ ("less than"), defined for some pairs of its elements, satisfying the following conditions:

1. If $u < v$, it is not true that $v < u$ or $u = v$.
2. If $u < v$ and $v < w$, then $u < w$.

The inequality $u < v$ will also be written $v > u$. If either $u < v$ or $u = v$, we shall write $u \leqslant v$ or $v \geqslant u$.

A set is said to be totally ordered, or a chain, if for any pair of elements u, v we have one of the relations $u < v$, $u = v$ or $u > v$.

We present a few examples of partially ordered sets arising frequently in analysis.

Finite- or infinite-dimensional space R_n of vectors with n real components (n is a natural number or infinity). Given two vectors $u = (u_1, \ldots, u_n)$ and $v = (v_1, \ldots, v_n)$, we shall say that $u < v$ if $u_i \leqslant v_i$ for all $i = 1, \ldots, n$ and $u_i < v_i$ for at least one i. The relation $u = v$ means that $u_i = v_i$ for all $i = 1, \ldots, n$.

The set of real-valued functions defined on a set P. Given two functions $u(t)$ and $v(t)$, we shall say that $u(t) < v(t)$ if $u(t) \leqslant v(t)$ for all $t \in P$ and $u(t) < v(t)$ for at least one point P.

Finite or infinite set of vectors $u(t) = \{u_1(t), \ldots, u_n(t)\}$ (n is a natural number or infinity) whose components are real functions $u_i(t)$ defined in domains P_i, respectively. Here $u(t) < v(t)$ if $u_i(t) \leqslant v_i(t)$ for all $i = 1, \ldots, n$ and $u_i(t) < v_i(t)$ for at least one i (in the sense of the previous definition).

Set of integral operators

$$Ax = \int_a^b K\,[t, s, x\,(s)]\, ds,$$

defined on some set Q of real-valued functions. We set $A < B$ if $Ax \leqslant Bx$ for all $x \in Q$ and $Ax < Bx$ for at least one $x \in Q$.

A subset M_1 of a partially ordered set M is said to be bounded above (below) if there is an element z such that $x \leqslant z$ ($x \geqslant z$) for all $x \in M_1$. The element z is then called an upper (lower) bound of M_1. It is readily shown that if a least upper (greatest lower) bound exists it is unique.

The least upper (greatest lower) bound of a nonempty set $M_1 \subset M$ is called its supremum (infimum) and denoted by $\sup M_1$ ($\inf M_1$).

The set of all elements x such that $a \leqslant x \leqslant b$ is called a segment and denoted by $[a, b]$.

2. *Set G.** Following [234], we denote by G a set with the following properties:

1. G is a partially ordered set with a smallest element θ, i.e., $\theta \leqslant x$ for every $x \in G$.

2. For any elements $u, v \in G$, there is a uniquely defined sum $u + v \in G$ such that

a) $u + v = v + u$ and $u + \theta = u$;

b) if $u, v, w \in G$ and $u \leqslant v$, then $u + w \leqslant v + w$;

c) if $u + v \leqslant w$, then $u \leqslant w$.

3. Any nonincreasing sequence $\{u_i\}$, where $u_i \in G$, i.e., a sequence such that $u_{i+1} \leqslant u_i$ for all $i = 1, 2, \ldots$, has a unique limit $u \in G$ (symbolically: $\lim_{i\to\infty} u_i = u$ or $u_i \searrow u$) with the following properties:

a) if $u_i = u$ for all $i = 1, 2, \ldots$, then $u_i \searrow u$;

b) if $u_i \searrow u$ and $v_i \searrow v$, then $u_i + v_i \searrow u + v$;

c) if $u_i \searrow u$, $v_i \searrow v$ and $u_i \leqslant v_i$, then $u \leqslant v$;

d) the limit of a sequence $\{u_i\}$ remains unchanged if finitely many initial elements of the sequence are changed.

Examples are the set of positive functions defined on a set M, the set of finite- or infinite-dimensional vectors with positive components, etc.

Let $\varphi(x)$ be a function defined on a set G, i.e. we have a rule which associates with each element x of G a well-defined element $\varphi(x)$ of G. We shall say that φ is nondecreasing if $u \leqslant v$ implies $\varphi(u) \leqslant \varphi(v)$ for any $u, v \in G$. The function φ is said to be monotonically continuous on a sequence $\{u_i\}$ if $u_i \searrow u$ implies $\varphi(u_i) \searrow \varphi(u)$. If a function is monotonically continuous on all sequences, we shall sometimes say that it is sequentially monotonically continuous.

Similar definitions may be introduced for functions of two or more variables. We shall say that a function $\varphi(u, v)$ is joinly nondecreasing in $u \in G$ and $v \in G$ if the relations $u \leqslant u_1$ and $v \leqslant v_1$ imply $\varphi(u, v) \leqslant \varphi(u_1, v_1)$. A function

---

**Translator's note.* As in [234], the letter G does not denote a particular set but any set with the desired properties.

of two variables is said to be nondecreasing in one of its variables if for any fixed value of one variable it is nondecreasing in the other, i.e., if $u, v, z \in G$ and $u \leqslant v$ then $\varphi(u, z) \leqslant \varphi(v, z)$, or if $u \leqslant v$ then $\varphi(z, u) \leqslant \varphi(z, v)$. Obviously, a function which is jointly nondecreasing is also nondecreasing in each variable separately. A function $\varphi(u, v)$ is said to be jointly monotonically continuous on sequences $\{u_i\}$ and $\{v_i\}$ if the relations $u_i \searrow u$ and $v_i \searrow v$ imply $\varphi(u_i, v_i) \searrow \varphi(u, v)$. Clearly, if $\varphi(u, v)$ is jointly monotonically continuous then it is also monotonically continuous in each variable separately.

3. *Functional inequality lemma.* A solution $x^* \in [a, b]$ of the equation $x = \varphi(x)$ is said to be upper (lower) on $[a, b]$ if any other solution $x \in [a, b]$ of the equation satisfies the inequality $x \leqslant x^*$ $(x^* \leqslant x)$.

LEMMA. *Let the function $\varphi(u)$, whose domain is a segment $\Delta = [\theta, k] \subset G$ and whose range is G, be nondecreasing and monotonically continuous on sequences, and let there exist a $b \in \Delta$ such that*

$$\varphi(b) \leqslant b. \tag{1.1}$$

Then the equation

$$u = \varphi(u) \tag{1.2}$$

has an upper solution $m(\varphi, b)$ on $[\theta, b]$. If moreover $\theta \leqslant p \leqslant b$ and

$$p \leqslant \varphi(p), \tag{1.3}$$

then $p \leqslant m(\varphi, b)$. If $\varphi(u) \leqslant \psi(u)$ for all $u \in [\theta, b]$, where ψ has the same properties as φ, and $\varphi(b) \leqslant b$, $\psi(b) \leqslant b$, then $m(\varphi, b) \leqslant m(\psi, b)$.

If the set G is such that any nondecreasing sequence $\{v_n\}$ $(v_n \in G)$ has a unique limit and the function $\varphi(u)$ is monotonically continuous on nondecreasing sequences, i.e., $v_n \nearrow v$ implies $\varphi(v_n) \nearrow \varphi(v)$, then equation (1.2) *also has a lower solution $n(\varphi, b)$ on $[\theta, b]$, and $n(\varphi, b) \leqslant n(\psi, b)$. Moreover, if $\varphi(q) \leqslant q$ $(q \in [\theta, b])$, then $n(\varphi, b) \leqslant q$.*

PROOF. Define a sequence $\{u_i\}$ $(i = 0, 1, 2, \ldots)$ recursively by

$$u_{i+1} = \varphi(u_i) \quad (i = 0, 1, 2, \ldots), \tag{1.4}$$

setting $u_0 = b$. We claim that $u_{i+1} \leqslant u_i$. Indeed, it follows from (1.4) by inequality (1.1) that $u_1 = \varphi(u_0) = \varphi(b) \leqslant b$. Suppose we have already proved that $u_i \leqslant u_{i-1}$. Since by assumption φ is nondecreasing, we have

$$u_{i+1} = \varphi(u_i) \leqslant \varphi(u_{i-1}) = u_i.$$

Hence it follows by induction that $u_{i+1} \leqslant u_i$ for all $i = 0, 1, 2, \ldots$. Since it is nonincreasing, the sequence $\{u_i\}$ has a unique limit u, and by the sequential monotone continuity of φ this limit satisfies (1.2). We claim that u is an upper solution, $u = m(\varphi, b)$. Let $\bar{u}$ be any solution of (1.2) on $[\theta, b]$. Then

$$\bar{u} = \varphi(\bar{u}) \leqslant \varphi(b) = u_1.$$

But if $\bar{u} < u_i$ for some i, then $\bar{u} \leqslant u_{i+1}$, since

$$\bar{u} = \varphi(\bar{u}) \leqslant \varphi(u_i) = u_{i+1}.$$

Thus $\bar{u} \leqslant u_i$ for all i. It follows that $\bar{u} \leqslant \lim_{i\to\infty} u_i = u$, so that u is indeed an upper solution of (1.2).

Since

$$\begin{gathered} p \leqslant \varphi(p) \leqslant \varphi(b) = u_1, \\ p \leqslant \varphi(p) \leqslant \varphi(u_1) = u_2, \\ \cdots\cdots\cdots\cdots \\ p \leqslant \varphi(p) \leqslant \varphi(u_i) = u_{i+1}, \end{gathered}$$

it follows that $p \leqslant \lim_{i\to\infty} u_i = m(\varphi, b)$.

The inequality $m(\varphi, b) \leqslant m(\psi, b)$ is true, since the elements u_i and v_i defined by

$$u_{i+1} = \varphi(u_i),\ v_{i+1} = \psi(v_i),\ u_0 = v_0 = b,$$

satisfy the condition $u_i \leqslant v_i$, as follows from the sequence of inequalities

$$\begin{gathered} u_1 = \varphi(b) \leqslant \psi(b) = v_1, \\ u_2 = \varphi(u_1) \leqslant \psi(u_1) \leqslant \psi(v_1) = v_2, \\ \cdots\cdots\cdots\cdots \\ u_i = \varphi(u_{i-1}) \leqslant \psi(u_{i-1}) \leqslant \psi(v_{i-1}) = v_i, \\ \cdots\cdots\cdots\cdots \end{gathered}$$

Similarly one proves the existence of a lower solution $n(\varphi, b)$ and the inequalities $n(\varphi, b) \leqslant n(\psi, b)$ and $n(\varphi, b) \leqslant q$. We need only set $u_0 = \theta$ in (1.4).

4. *Space R metrized by elements of a set G.* We first define the general concept of an abstract generalized metric space. An arbitrary set E of elements $x, y, \ldots$ is called an abstract space metrized by elements of a set G if for each pair of elements x, y in E there is a well-defined element $d(x, y)$ of G, called the distance between x and y, possessing the following properties:

1. $d(x, y) = 0$ if and only if $x = y$.
2. $d(x, y) = d(y, x)$ for any x and y in E.
3. $d(x, y) \leqslant d(x, z) + d(z, y)$ for any elements x, y, z in E (triangle inequality).

The set of all elements x in E such that $d(x, x^*) \leqslant r$ $(r \in G)$ is called the ball with center x^* and radius r. We denote this ball by $S(x^*, r)$.

We let R denote a space with the following properties [234]:

1. R is an abstract space metrized by elements of G.

2. For some sequences $\{x_i\}$, $x_i \in R$, there is a uniquely defined limit $\lim_{n\to\infty} x_n = x$ $(x \in R)$ which is not affected by changing finitely many initial terms of the sequence and such that if $x_i = s \in R$, then $\lim_{i\to\infty} x_i = s$.

3. Any ball $S(x^*, r)$, $x^* \in R$, $r \in G$, is a closed set in the sense of the convergence just defined on R.

4. The space R is complete, in the sense that if $\{x_n\}$ satisfies the condition

$$d(x_n, x_{n+m}) \leqslant c_n \quad (n, m = 0, 1, 2, \ldots),$$

where $c_n \searrow 0$, then $\{x_n\}$ converges to some element $x \in R$.

Let $f(x)$ be a function defined on some subset D of an abstract metric space E, with values in E. We shall say that f satisfies a generalized Lipschitz condition in D if for all $x, y \in D$

$$d(f(x), f(y)) \leqslant \varphi(d(x, y)), \tag{1.5}$$

where φ is a function defined on G and taking values in G, which we call a Lipschitz function or Lipschitz operator.

Similarly one defines a generalized Lipschitz condition for functions of two variables:

$$d(f(x, y), f(z, t)) \leqslant \varphi(d(x, z), d(y, t)) \quad (x, y, z, t \in E), \tag{1.6}$$

where φ, called a Lipschitz function or operator, is defined in $G \times G$ and has values in G.

5. *Linear systems and K-lineals.* A set E is called a real linear system if it possesses the following properties:

1. For any two elements $x, y \in E$, there is a uniquely defined sum $x + y \in E$, and the following conditions hold:

a) $x + y = y + x$ (commutativity of addition);

b) $(x + y) + z = x + (y + z)$ (associativity of addition).

2. For any $x \in E$ and any real number (scalar) λ, there is a uniquely defined product $\lambda x \in E$ such that

a) $\lambda(\mu x) = (\lambda\mu)x$, where μ is a real number;

b) $1 \cdot x = x$;

c) $0x = \theta$, where $\theta \in E$ is the zero element.

3. Addition and multiplication by scalars satisfy the following conditions:

a) $(\lambda + \mu)x = \lambda x + \mu x$;

b) $\lambda(x + y) = \lambda x + \lambda y$.

Note that the zero element θ is uniquely defined by the property that $x + \theta = x$ for any $x \in E$.

Examples of real linear systems are provided by the sets considered in Subsection 1, with the usual definitions of addition of vectors and functions and multiplication by numbers.

Suppose that in some real linear system E we have a special class of positive elements, i.e., it is somehow stipulated which elements $x \in E$ are to be considered greater than the zero element θ $(x > \theta)$. The set of all these elements, together with θ, constitutes the class of positive elements E_+.

We now endow the set E with a partial order, setting $x \geqslant y$ if $x - y \geqslant \theta$. The set of all $x \in E$ such that $x \leqslant \theta$ forms the class of negative elements E_-.

A real linear system will be called a K-lineal (Kantorovič lineal) if it contains a class of positive elements and the partial order thereby defined satisfies the following conditions:

1. If $x > \theta$, then $x \neq \theta$.
2. If $x > \theta$ and $y > \theta$, then $x + y > \theta$.
3. Any elements x and y have a supremum $\sup(x, y)$.
4. If $x > \theta$ and $\lambda > 0$, then $\lambda x > \theta$.

It can be shown that under these conditions the infimum $\inf(x, y)$ of any two elements x and y also exists.

Given a K-lineal E, we define the modulus $|x|$ of an element x by

$$|x| = x_+ + x_-,$$

where

$$x_+ = \sup(x, \theta), \quad x_- = \sup(-x, \theta).$$

The element x_+ is called the positive part of x and x_- its negative part.

The modulus thus defined has the following properties:

1) $|x| = \theta$ if and only if $x = \theta$;

2) $|\lambda x| = |\lambda|\,|x|$ for any real number λ;

3) $|x + y| \leqslant |x| + |y|$.

Each element x of a K-lineal may be expressed as $x = x_+ - x_-$.

A K-lineal is said to be archimedean if, whenever $nx \leqslant y$ for some $x, y \in K$ and any natural number n, then $x \leqslant \theta$.

We shall say that a sequence $\{x_n\}$, $x_n \in E$, converges uniformly to $x \in E$ if there exist an element $b \in E$ and a monotonically decreasing sequence of numbers $\{\lambda_n\}$, converging to zero, such that $|x_n - x| \leqslant \lambda_n b$, $n = 1, 2, \ldots$.

It is not hard to prove the following statements [168]:

1) If $a_n = a$, then $\lim a_n = a$.

2) If a_n converges to a limit a, then any subsequence a_{k_n} converges to the same limit a.

3) If $\lim a_n = a$ and $\lim b_n = b$, then $\lim(a_n + b_n)$ exists and is equal to $a + b$.

4) If $\lim a_n = a$ and $\lim \alpha_n = \alpha$ (where α and the α_n are real numbers), then $\lim \alpha_n a_n$ exists and is equal to αa.

5) If $\lim a_n = \theta$ and $\theta \leqslant b_n \leqslant a_n$, then $\lim b_n = \theta$.

In addition, if the K-lineal is archimedean, then:

6) If $\lim a_n = a$, $a_n \geqslant \theta$, then $a \geqslant \theta$.

It can also be shown that a sequence converges to a unique limit.

Examples of archimedean K-lineals are the partially ordered sets considered in subsection 1, and also the following:

1) The K-lineal $l^{(p)}$ $(p > 0)$: the set of all real number sequences $x = \{\xi_n\}$ satisfying the condition

$$\sum_{n=1}^{\infty} |\xi_n|^p < +\infty.$$

An element $x = (\xi_1, \xi_2, \dots)$ is positive if $\xi_n \geqslant 0$ for all $n = 1, 2, \dots$.

2) The K-lineal L^p $(p > 0)$: the set of all real measurable functions $x(t)$ defined on $[a, b]$ and satisfying the condition

$$\int_b^a |x(t)|^p \, dt < +\infty.$$

Here we define $x > \theta$ if $x(t) \geqslant 0$ almost everywhere and $x(t) > 0$ on a set of positive measure.

We now consider the concept of a positive linear operator defined in a K-lineal.

An operator A from some space E to a space E_1 is said to be linear if, for all $x, y \in E$ and real λ, μ,

$$A(\lambda x + \mu y) = \lambda Ax + \mu Ay.$$

A linear operator A from a K-lineal E to E is said to be positive linear if the inequality $a \geqslant \theta$ implies that $Aa \geqslant \theta$. It can be proved [168] that a positive linear operator is continuous, i.e., if $\lim x_n = x$ then $\lim Ax_n = Ax$ in the sense of uniform convergence. Moreover, if A is a positive linear operator such that $\Sigma_0^\infty A^i a$ is uniformly convergent (in other words, the sequence $\Sigma_0^n A^i a$ is uniformly convergent) for any a in the K-lineal, then the inverse operator $(I - A)^{-1}$ exists, i.e., an operator such that $(I - A)^{-1}(I - A) = I$ (where I is the identity operator), and then $(I - A)^{-1} a = \Sigma_0^\infty A^i a$, and $(I - A)^{-1}$ is positive linear and therefore also continuous [223].

6. *Lattice-normed spaces.* A space is said to be lattice-normed by the elements of a K-lineal N, or simply N-normed, if it is a real linear system in which for each element $x \in E$ there is a positive element $\|x\| \in N$, called the generalized norm of x, such that

1. $\|x\| \geqslant \theta$, and $\|x\| = \theta$ only if $x = \theta$;

2. $\|\lambda x\| = |\lambda| \, \|x\|$ (where λ is a real number);
3. $\|x + y\| \leqslant \|x\| + \|y\|$.

The concept of lattice-normed spaces is due to Kantorovič.

The principal type of convergence in an N-normed space is convergence in norm: a sequence $\{x_n\}$, $x_n \in E$, converges to an element $x \in E$ if $\lim \|x_n - x\| = \theta$ in the sense of uniform convergence. This convergence concept has the usual properties of convergence in normed spaces; in particular, the limit is unique.

A sequence $\{x_n\}$ of elements of an N-normed space E will be called a Cauchy sequence if $\lim_{m,n\to\infty} \|x_m - x_n\| = \theta$ in the sense of uniform convergence. If every Cauchy sequence of elements of E has a limit in E, we shall call E a complete space.

§2. General error estimates for approximate solutions of operator equations

Consider the (not necessarily linear) operator equation

$$x = Tx, \tag{2.1}$$

where T is an operator with domain and range in a space R metrized by the elements of a set G (see §1).

1. Let x^* be an exact solution of equation (2.1) and v some approximate solution, or in general some given element of a space E; assume that both x^* and v lie in some ball $S(z, r)$. We wish to find an upper bound for the distance of v from x^*, i.e., to estimate $d(x^*, v)$.

THEOREM 2.1. *Suppose that for all x in the ball $S(z, r)$*

$$d(Tx, Tv) \leqslant a(d(x, v)), \tag{2.2}$$

where $a(d)$ is a function, nondecreasing and sequentially monotonically continuous, defined on some segment $[\theta, k] \subset G$, taking values in G and generally depending on v, and let

$$d(Tv, v) \leqslant q \leqslant b, \quad 2r \leqslant b \quad (b \in [\theta, k]), \qquad \varphi(b) \equiv q + a(b) \leqslant b. \tag{2.3}$$

Then

$$d(x^*, v) \leqslant m(\varphi, b), \tag{2.4}$$

where $m(\varphi, b)$ is an upper solution of the equation

$$x = \varphi(x) \tag{2.5}$$

on $[\theta, b]$.

PROOF. Since x^* is a solution of (2.1), it follows from the triangle inequality and condition (2.2) that

$$d(x^*, v) = d(Tx^*, v) \leqslant d(Tx^*, Tv) + d(Tv, v)$$
$$\leqslant a(d(x^*, v)) + q = \varphi(d(x^*, v)).$$

Thus the element $d(x^*, v)$ has the same properties as the element p in the lemma of §1, and so estimate (2.4) is valid.

REMARK 1. If we replace the ball $S(z, r)$ by the ball $S(v, r)$ of radius r centered at v, the condition $2r \leqslant b$ of Theorem 2.1 may be replaced by the less restrictive condition $r \leqslant b$.

2. The estimate (2.4) was proved on the assumption that there is a solution of equation (2.1) in the ball $S(z, r)$. We shall now prove a theorem which furnishes sufficient conditions for the existence of a unique solution of (2.1) in a prescribed ball $S(z, r)$. The solution will be the limit of a sequence of approximate solutions determined by the familiar method of successive approximations.

THEOREM 2.2. *Suppose that any two elements $x, y \in S(z, r)$ satisfy a generalized Lipschitz condition*

$$d(Tx, Ty) \leqslant A(d(x, y)), \tag{2.6}$$

where $A(d)$ is a nondecreasing function, sequentially monotonically continuous, with domain in some segment $[\theta, k] \subset G$ and range in G, such that the equation $x = Ax$ has exactly one solution, θ, on $[\theta, k]$, and there exists $q \in [\theta, k]$ such that $d(z, Tz) \leqslant q$ and

$$q + A(r) \leqslant r, \tag{2.7}$$

$A(b) \leqslant b$ if $2r \leqslant b$.

Then equation (2.1) *has a unique solution in the ball $S(z, r)$, which is the limit of the sequence $\{x_n\}$ defined recursively by*

$$x_n = Tx_{n-1} \quad (n = 1, 2, \ldots) \tag{2.8}$$

(x_0 is an arbitrary element of $S(z, r)$).

PROOF. We first show that under the above assumptions the operator T maps $S(z, r)$ into itself. Indeed, if $x \in S(z, r)$ it follows from (2.6) and (2.7) that

$$d(Tx, z) \leqslant d(Tx, Tz) + d(Tz, z) \leqslant A(d(x, z)) + q \leqslant A(r) + q \leqslant r,$$

so that $Tx \in S(z, r)$. Since $x_0 \in S(z, r)$, this implies that all the x_n $(n = 0, 1, 2, \ldots)$ are in $S(z, r)$, i.e., $d(z, x_m) \leqslant r$. Next, we have

$$d(x_0, x_m) \leqslant d(x_0, z) + d(z, x_m) \leqslant r + r = 2r \leqslant b.$$

Assuming that

$$d(x_n, x_{n+m}) \leqslant A_n(b), \tag{2.9}$$

where the A_n are defined by

$$A_n(b) = A(A_{n-1}(b)), \quad A_0(b) = b, \tag{2.10}$$

we see that

$$d(x_{n+1}, x_{m+n+1}) = d(Tx_n, Tx_{m+n}) \leqslant A(d(x_n, x_{m+n})) \leqslant A(A_n(b)) = A_{n+1}(b).$$

By induction, therefore,

$$d(x_n, x_{n+m}) \leqslant A_n(b) \tag{2.11}$$

for any $m, n = 0, 1, 2, \ldots$. It is readily seen that the sequence $\{A_n(b)\}$ is nondecreasing and bounded below by θ, so that it has a limit c which, by (2.10), must satisfy the equation $x = Ax$. But since this equation has by assumption a unique solution θ, the nondecreasing sequence $\{A_n(b)\}$ converges to θ, and this means (by the completeness of R) that the sequence $\{x_n\}$ converges to some limit $x \in R$. Since $S(z, r)$ is closed, it follows that $x \in S(z, r)$. Letting $m \longrightarrow \infty$ in (2.11), we see that

$$d(x_n, x) \leqslant A_n(b). \tag{2.12}$$

We now show that the limit x is a solution of equation (2.1). Indeed,

$$d(x, Tx) \leqslant d(x, x_n) + d(x_n, Tx) = d(x, x_n) + d(Tx_{n-1}, Tx)$$
$$\leqslant d(x, x_n) + A(d(x, x_{n-1})) \leqslant A_n(b) + A(A_{n-1}(b)) = A_n(b) + A_n(b),$$

and so $d(x, Tx) = \theta$, i.e., $x = Tx$.

It is readily seen that x is the unique solution of (2.1) in $S(z, r)$. Supposing the contrary, let $\bar{x}$ be another solution in $S(z, r)$. Then

$$d(x, x_0) \leqslant d(x, z) + d(z, x_0) \leqslant 2r \leqslant b,$$

and since the inequality $d(\bar{x}, x_n) \leqslant A_n(b)$ implies

$$d(\bar{x}, x_{n+1}) = d(T\bar{x}, Tx_n) \leqslant A(d(\bar{x}, x_n)) \leqslant A(A_n(b)) = A_{n+1}(b),$$

it follows by induction that

$$d(\bar{x}, x_n) \leqslant A_n(b) \tag{2.13}$$

for all $n = 0, 1, 2, \ldots$. Then

$$d(x, x) \leqslant d(x, x_n) + d(x, x_n) \leqslant A_n(b) + A_n(b),$$

whence $d(x, \bar{x}) = \theta$, so that $x = \bar{x}$.

REMARK 2. For the function $a(c)$ in (2.2) we may clearly take the function $A(c)$ defined by (2.6). However, we have assumed that (2.6) is true for all

x and y in $S(z, r)$, whereas (2.2) holds only for all x in $S(z, r)$ and a fixed element $v \in S(z, r)$. Thus $a(c)$ may always be chosen in such a way that $a(c) \leqslant A(c)$ for $c \in [\theta, b]$.

REMARK 3. If the initial element in the definition of the sequence $\{x_n\}$ is taken to be the center of $S(z, r)$, $x_0 = z$, the stipulation in Theorem 2.2 that $A(b) \leqslant b$ for $2r \leqslant b$ may be dropped. Theorem 2.2 then reduces to a theorem proved in [105].

3. We now assume that equation (2.1) is given in a space E lattice-normed by an archimedean K-lineal N, and its exact solution x^* and approximate solution v lie in some subset D of E.

THEOREM 2.3. *Suppose that for all x in a subset D of an N-normed space E*

$$\|Tx - Tv\| \leqslant L\|x - v\| \quad (v \in D), \tag{2.14}$$

where L is a positive linear operator with domain and range in N, such that $\Sigma_0^\infty L^i a$ is uniformly convergent for any $a \in N$. Then

$$\|x^* - v\| \leqslant (I - L)^{-1}\|Tv - v\|. \tag{2.15}$$

PROOF. By virtue of (2.14), it follows from the equality $x^* = Tx^*$ that

$$\|x^* - v\| = \|Tx^* - v\| \leqslant \|Tx^* - Tv\| + \|Tv - v\|$$
$$\leqslant L\|x^* - v\| + \|Tv - v\|,$$

whence, in view of the fact that the uniform convergence of $\Sigma_0^\infty L^i a$ for any a and positive L implies the existence of a positive linear operator $(I - L)^{-1}$ (see §1), we obtain the estimate (2.15).

4. The operators a and L in inequalities (2.2) and (2.14) depend essentially on the domain D in which the inequality holds. The dependence is such that if a_1 and L_1 are the appropriate operators for a domain D_1, and a and L correspond to a domain D such that $D_1 \subset D$, then we may clearly assume that $a_1(c) \leqslant a(c)$ and $L_1 c \leqslant Lc$ for any c in the domains of definition of a_1 and L_1. This observation is the basis for the following procedure, aimed at sharpening the estimates (2.4) and (2.15).

Since the exact solution x^* of equation (2.1) lies in the domain D and in the ball S_1 about v with radius $r_1 = m(\varphi_1, b)$, it is in their intersection $D_1 = D \cap S_1$, which also contains the element v. We may therefore replace the function a in (2.2) by a function a_1 corresponding to the smaller domain $D_1 \subset D$. Then the solution x^* of equation (2.1) corresponding to the domain D_1 and the ball S_2 about v with radius $r_2 = m(\varphi_1, b)$, where $\varphi_1(b) = q + a_1(b)$, is in the intersection $D_2 = D_1 \cap S_2$, which again also contains v. Thus we may replace a_1 by a new function a_2 corresponding to the (generally) smaller domain $D_2 \subset$

D_1. Continuing in this manner, we obtain a sequence of domains $D_{i+1} \subset D_i$ $(i = 1, 2, \ldots)$, each containing both x^* and v. But since $D_{i+1} \subset D_i$ we may assume that $a_{i+1}(c) \leqslant a_i(c)$ for any c in the domain of a_{i+1}. Hence

$$\varphi_{i+1}(b) = q + a_{i+1}(b) \leqslant q + a_i(b) = \varphi_i(b),$$

and by the lemma in §1 we have

$$m(\varphi_{i+1}, b) \leqslant m(\varphi_i, b),$$

so that the estimate (2.2) is generally sharpened.

Analogous arguments go through for the estimate (2.14).

§3. Nonstationary iterative method for equations in abstract metric and normed spaces

Let E be an abstract space. Consider the (not necessarily linear) operator equation

$$x = Tx, \tag{3.1}$$

where T is an operator from E to E; x is an unknown element of E.

In this section we shall investigate convergence criteria and error estimates for a nonstationary iterative method whose general scheme is prescribed by the recurrence relation

$$x_n = T_n x_{n-1} \quad (n = 1, 2, \ldots), \tag{3.2}$$

where the T_n are operators from E to E and x_0 is some element of E.

The iterative method (3.2) was considered in [50] and [166], under different assumptions, for the case of an abstract space R metrized by elements of a certain semiordered lineal.

1. Let E be an abstract space R metrized by the elements of a set G (see §1); suppose that equation (3.1) has a solution x in some set D, all approximations x_n also belonging to D.

THEOREM 3.1. *Suppose that for all $x, y \in D$*

$$d(T_n x, T_n y) \leqslant \varphi_n(d(x, y)), \tag{3.3}$$

where $\varphi_n(d)$ are functions defined on some segment $\Delta = [\theta, k] \subset G$ with values in G. Suppose in addition that

$$d(T_n x, Tx) \leqslant a_n \ (a_n \searrow \theta), \tag{3.4}$$

if x is a solution of equation (3.1), *and let*

$$\varphi(b) + a_1 \leqslant b, \tag{3.5}$$

where $b \in [\theta, k]$ is the diameter of D and $\varphi(c)$ is a nondecreasing sequentially monotonically continuous function such that $\varphi_n(c) \leqslant \varphi(c)$ for all $c \in [\theta, b]$ and the equation $x = \varphi(x)$ has a unique solution $x = \theta$.

Then x is the unique solution of equation (3.1) in D, and the sequence $\{x_n\}$ defined by (3.2) converges to x for any $x_0 \in D$, in such a way that

$$d(x_n, x) \leqslant \varphi_n(d(x_{n-1}, x)) + a_n, \tag{3.6}$$

$$d(x_n, x) \leqslant A_n(b), \tag{3.7}$$

where $A_n(b)$ is defined by

$$A_n(b) = \varphi(A_{n-1}(b)) + a_n, \quad A_0(b) = b \tag{3.8}$$

With this definition, $A_n(b) \searrow \theta$ as $n \longrightarrow \infty$.

PROOF. The error estimate (3.6) follows directly from the equalities $x = Tx$ and $x_n = T_n x_{n-1}$ by virtue of condition (3.3):

$$\begin{aligned} d(x_n, x) = d(T_n x_{n-1}, Tx) &\leqslant d(T_n x_{n-1}, T_n x) + d(T_n x, Tx) \\ &\leqslant \varphi_n(d(x_{n-1}, x)) + a_n. \end{aligned} \tag{3.9}$$

Since $x_n \in D$ for all $n = 0, 1, 2, \ldots$, it follows that $d(x_n, x) \leqslant b$. The conditions $\varphi_n(c) \leqslant \varphi(c)$ ($c \in [\theta, b]$) and (3.5) imply

$$\begin{aligned} d(x_1, x) \leqslant \varphi_1(d(x_0, x)) + a_1 &\leqslant \varphi(d(x_0, x)) + a_1 \\ &\leqslant \varphi(b) + a_1 = A_1(b) \leqslant b = A_0(b). \end{aligned}$$

Assuming that

$$A_n(b) \leqslant A_{n-1}(b) \tag{3.10}$$

for some n, we deduce from (3.4) that

$$A_{n+1}(b) = \varphi(A_n(b)) + a_{n+1} \leqslant \varphi(A_{n-1}(b)) + a_n = A_n(b),$$

and so by induction inequality (3.10) is true for any $n = 1, 2, \ldots$.

But $A_n(b)$ is a nonincreasing sequence of elements of G, and therefore it has a limit c which, by (3.8) and the sequential monotone continuity of φ, must satisfy the equation $x = \varphi(x)$. Thus $c = \theta$, for by assumption this equation has no nontrivial solution.

The uniqueness of the solution is readily proved by reductio ad absurdum. In fact, were there another solution $\bar{x}$ of (3.1) in D, it would follow from the above arguments that $d(x_n, \bar{x}) \leqslant A_n(b)$, and then

$$d(x, \bar{x}) \leqslant d(x, x_n) + d(\bar{x}, x_n) \leqslant A_n(b) + A_n(b),$$

which is possible only if $d(x, \bar{x}) = \theta$, i.e., $x = \bar{x}$.

REMARK 1. In general it is quite difficult to find estimates a_n for the distances $d(T_n x, Tx)$, since the solution x of (3.1) is unknown. However, if the operator T_n have the property that for each n the equation

$$x = T_n x \tag{3.11}$$

is equivalent to (3.1), then $d(T_n x, Tx)$ is of course always equal to θ, and we may thus put $a_n = \theta$ for all $n = 1, 2, \ldots$.

REMARK 2. Let us replace G by a set G_1 with the following additional property: for certain (not necessarily nonincreasing) sequences $\{u_n\}$, $u_n \in G$, we have a uniquely defined limit θ ($\lim u_n = \theta$), independent of modification of finitely many initial terms in the sequences, such that if $\lim u_n = \theta$ and $\lim v_n = \theta$, then $\lim(u_n + v_n) = \theta$, and if $\theta \leqslant a_n \leqslant b_n$ and $\lim b_n = \theta$, then $\lim a_n$ exists and is equal to θ. Suppose, moreover, that the functions $\varphi_n(d)$ defined by (3.3) satisfy (instead of the previous conditions)

$$\lim \varphi_n(c_n) = \theta, \quad \text{if} \quad \lim c_n = \theta.$$

Then

$$\lim d(T_n x, Tx) = \theta \tag{3.12}$$

is a necessary condition for convergence of the sequence $\{x_n\}$ to a solution x of equation (3.1).

Indeed, since

$$\theta \leqslant d(T_n x, Tx) \leqslant d(T_n x, T_n x_{n-1}) + d(T_n x_{n-1}, Tx)$$
$$\leqslant \varphi_n(d(x, x_{n-1})) + a(x_n, x),$$

it follows from the above assumptions that $\lim d(x_n, x) = \theta$ implies (3.12).

2. Let E be an abstract space normed by the elements of an archimedean K-lineal N. We consider the case that equations (3.1) and (3.11) are equivalent for every n on some closed subset D of E.

THEOREM 3.2. *Let equation* (3.1) *have a solution $x \in D$, suppose that $x_n \in D$ for all n and for all $x, y \in D$ let*

$$\| T_n x - T_n y \| \leqslant M_n \| x - y \| \quad (n = 1, 2, \ldots), \tag{3.13}$$

where M_n are positive linear operators from N to N such that the sequence

$$\prod_{i=1}^{n} M_i a = M_n M_{n-1} \ldots M_1 a$$

converges uniformly to θ for any $a \in N$.

Then equation (3.1) *has a unique solution x on D, and the sequence $\{x_n\}$*

defined by (3.2) *converges to x for any* $x_0 \in D$.

If $(1 - M_n)^{-1}$ *exists, an error estimate for the nth approximation is given by*

$$\|x - x_n\| \leqslant (I - M_n)^{-1} M_n \|x_n - x_{n-1}\|. \tag{3.14}$$

PROOF. Since x is a solution of (3.11), we have

$$x - x_n = T_n x - T_n x_{n-1} = T_n x - T_n x_n + T_n x_n - T_n x_{n-1},$$

whence, by (3.13),

$$\|x - x_n\| \leqslant M_n \|x - x_n\| + M_n \|x_n - x_{n-1}\|$$

and the existence of a positive inverse $(1 - M_n)^{-1}$ implies (3.14).

The equality $x - x_n = T_n x - T_n x_{n-1}$ also yields

$$\theta \leqslant \|x - x_n\| \leqslant M_n \|x - x_{n-1}\| \leqslant \prod_{i=1}^{n} M_i \|x - x_0\|$$

and since $\Pi_1^n M_i a$ is uniformly convergent for any $a \in N$ we have $\lim \|x - x_n\| = \theta$.

The uniqueness of the solution x of equation (3.1) in D is again proved indirectly. If there were another solution $\bar{x} \in D$, it would follow that $x - \bar{x} = T_n x - T_n \bar{x}$ $(n = 1, 2, \ldots)$, and by (3.13)

$$\|x - \bar{x}\| \leqslant \prod_{i=1}^{n} M_i \|x - \bar{x}\|,$$

and the uniform convergence of $\Pi_1^n M_i a$ to θ for any a implies that this is possible only if $\|x - \bar{x}\| = \theta$, or $x = \bar{x}$.

§4. Nonstationary generalized iterative method for equations in a space R metrized by elements of a set G

1. As in the preceding section, we consider a (not necessarily linear) operator equation

$$x = Tx, \tag{4.1}$$

where T is an abstract function with domain and range in a space R metrized by the elements of a set G (see §1).

Let (4.1) have a solution x in a closed subset D of R.

In this section we investigate convergence criteria and error estimates for a generalized iterative method, in which the successive approximations x_n to the solution of (4.1) are defined as solutions of operator equations

$$x_n = T_n(x_n, x_{n-1}) \quad (n = 1, 2, \ldots), \tag{4.2}$$

where the T_n are operators depending on two arguments, from $R \times R$ to R, i.e., $T_n(x, y) \in R$ for $x, y \in R$ and x_0 is an arbitrary element of R.

In a somewhat more general form, allowing the operators T_n to depend on the elements $x_0, x_1, \ldots, x_{n-2}$, the algorithm (4.2) was considered in [231] for operator equations (4.1) in a Banach space, and for operator equations in a space lattice-normed by the elements of an archimedean K-lineal in [168].

The operators T_n are usually chosen in such a way that it is easier to solve (4.2) than the original equation (4.1).

We shall assume that

$$d(T_n(x, x),\ Tx) \leqslant a_n\,(a_n \searrow \theta), \tag{4.3}$$

where x is a solution of (4.1).

THEOREM 4.1. *Suppose that for all $x, y, z, t \in D$*

$$d(T_n(x, y), T_n(z, t)) \leqslant \varphi_n(d(x, z), d(y, t)), \tag{4.4}$$

where $\varphi_n(c, d)$ is a function defined on $\Delta \times \Delta$, $\Delta = [\theta, k] \subset G$, with values in the set G, and that $\varphi_n(c, d)$ is nondecreasing in its first argument and sequentially monotonically continuous.

Assume moreover that condition (4.3) holds and equations (4.2) are solvable for x_n in the set D, $n = 1, 2, \ldots$, and in addition

$$\varphi(b, b) + a_1 \leqslant b, \tag{4.5}$$

where b is the diameter of D and $\varphi(c, d)$ is a nondecreasing function of both arguments defined on $[\theta, b] \times [\theta, b]$ with values in G, jointly sequentially monotonically continuous and such that

$$\varphi_n(c, d) \leqslant \varphi(c, d) \tag{4.6}$$

for $c, d \in [\theta, b]$ and such that the equation $x = \varphi(x, x)$ has a unique solution $x = \theta$.

Then equation (4.1) has a unique solution x in D, which is the limit of the sequence of solutions x_n of equations (4.2) in D, for any $x_0 \in D$, and we have the following error estimates:

a) $d(x_n, x) \leqslant B_n(b) \quad (n = 1, 2, \ldots),$ (4.7)

where $B_n(b)$ is an upper solution on $[\theta, b]$ of the equation

$$u = \varphi(u, B_{n-1}(b)) + a_n, \quad B_0(b) = b \quad (n = 1, 2, \ldots), \tag{4.8}$$

and $B_n(b) \searrow \theta$ as $n \to \infty$.

b) $d(x_n, x) \leqslant m(\psi_n, b),$ (4.9)

where $m(\psi_n, b)$ is an upper solution on $[\theta, b]$ of the equation

$$u = \psi_n(u) = \varphi_n(u, d(x_{n-1}, x)) + a_n. \tag{4.10}$$

PROOF. Since $x \in D$ and $x_n \in D$ for all $n = 1, 2, \ldots$, it follows from the equalities $x = Tx$ and $x_n = T_n(x_n, x_{n-1})$ via condition (4.4) that

$$\begin{aligned} d(x_n, x) &= d(T_n(x_n, x_{n-1}), Tx) \\ &\leqslant d(T_n(x_n, x_{n-1}), T_n(x, x)) + d(T_n(x, x), Tx) \\ &\leqslant \varphi_n(d(x_n, x), d(x_{n-1}, x)) + a_n, \end{aligned} \tag{4.11}$$

and since $d(x_n, x) \leqslant b$ for all $n = 0, 1, 2, \ldots$, it follows from (4.6), (4.3) and (4.5) that

$$d(x_n, x) \leqslant \varphi(d(x_n, x), d(x_{n-1}, x)) + a_n \leqslant \varphi(b, b) + a_1 \leqslant b. \tag{4.12}$$

Hence, for $n = 1$,

$$d(x_1, x) \leqslant \varphi(d(x_1, x), b) + a_1 \leqslant b,$$

and by the lemma in §1 on functional inequality

$$d(x_1, x) \leqslant B_1(b) \leqslant b = B_0(b).$$

Assuming that

$$d(x_n, x) \leqslant B_n(b) \leqslant b \tag{4.13}$$

for some n, we deduce from (4.12) that

$$\begin{aligned} d(x_{n+1}, x) &\leqslant \varphi(d(x_{n+1}, x), d(x_n, x)) + a_{n+1} \\ &\leqslant \varphi(d(x_{n+1}, x), B_n(b)) + a_{n+1} \leqslant b, \end{aligned}$$

whence, by the same lemma,

$$d(x_{n+1}, x) \leqslant B_{n+1}(b) \leqslant b.$$

Thus, by induction, (4.13) is true for all n.

We now show that $B_n(b) \searrow \theta$. We have already proved that $B_1(b) \leqslant B_0(b)$. Suppose that for some n

$$B_n(b) \leqslant B_{n-1}(b). \tag{4.14}$$

Since $\varphi(c, d)$ is nondecreasing in both arguments and $a_n \leqslant a_{n-1}$,

$$\bar{\psi}_n(u) = \varphi(u, B_n(b)) + a_n \leqslant \varphi(u, B_{n-1}(b)) + a_{n-1} = \bar{\psi}_{n-1}(u)$$

for any $u \in [\theta, b]$. Then, by the functional inequality lemma,

$$m(\overline{\psi}_n, b) \leqslant m(\overline{\psi}_{n-1}, b),$$

i.e., $B_{n+1}(b) \leqslant B_n(b)$, and so (4.14) holds for any n.

Since $\{B_n(b)\}$ is a nonincreasing sequence of elements of G, bounded below by θ, it has a limit c. Letting $n \longrightarrow \infty$ in the equality

$$B_n(b) = \varphi(B_n(b),\ B_{n-1}(b)) + a_n,$$

we get $c = \varphi(c, c)$, and by our assumptions on φ this implies that $c = \theta$.

Estimate (4.9) may be proved as follows. By (4.11),

$$d(x_n, x) \leqslant \psi_n(d(x_n, x)) = \varphi_n(d(x_n, x),\ d(x_{n-1}, x)) + a_n$$
$$\leqslant \varphi(d(x_n, x),\ d(x_{n-1}, x)) + a_n \leqslant \varphi(b, b) + a_1 \leqslant b,$$

whence, by the functional inequality lemma, we obtain (4.9).

The uniqueness of the solution $x \in D$ of (4.1) is proved in exactly the same way as in Theorem 3.1, with $A(b)$ replaced by $B(b)$.

REMARK 1. In the general case, it is difficult to find estimates for the quantities $d(T_n(x, x), Tx)$ and $d(T_n x, Tx)$ in §3, since x is unknown. In certain cases estimates can nevertheless be found. For example, if the equations

$$x = T_n(x, x) \tag{4.15}$$

are equivalent to (4.1) for each n, then $d(T_n(x, x), Tx) = \theta$.

REMARK 2. If we replace G by the set G_1 as in §3 (Remark 2) and the function $\varphi_n(c, d)$ in (4.4) has the property

$$\lim \varphi_n(c_n, d_n) = \theta, \quad \text{if} \quad \lim c_n = \theta,\ \lim d_n = \theta,$$

then the condition

$$\lim d(T_n(x, x), Tx) = \theta, \tag{4.16}$$

where x is a solution of (4.1), is a necessary condition for the sequence $\{x_n\}$ ($x_n \in D$) to converge to the solution $x \in D$ of (4.1). Indeed, since

$$\theta \leqslant d(T_n(x, x), Tx) \leqslant d(T_n(x, x), T_n(x_n, x_{n-1}))$$
$$+ d(T_n(x_n, x_{n-1}), Tx) \leqslant \varphi_n(d(x, x_n),\ d(x, x_{n-1})) + d(x_n, x),$$

it follows from our assumptions concerning $\varphi_n(c, d)$ that (4.16) is a consequence of $\lim d(x_n, x) = \theta$.

2. The rate of convergence of the successive approximations (4.2) depends essentially on the choice of the operators T_n. If equations (4.1) and (4.15) are equivalent, it is quite natural to choose the T_n in such a way that the principal part of $T_n(x, y)$ is a function of x alone, i.e., so that this expression depends strongly on x and only weakly on y. In that case the equations (4.2) for x_n will

be in a certain sense close to the original equation (4.1) for x.

The following construction of the operators T_n is frequently very effective. Suppose that on some set D equation (4.1) is equivalent to the sequence of equations

$$x = F_n(x, x) \quad (n = 1, 2, \ldots), \tag{4.17}$$

where $F_n(x, y) \in E$ (E is some abstract space) for $x, y \in D$ ($D \subset E$). This means that all solutions of (4.1) in D are also solutions of each of equations (4.17) in D, and conversely.

It is clear that all solutions of (4.17) for fixed n also satisfy

$$x = F_{n,k}(x, x) \tag{4.18}$$

for the same n, where the operators $F_{n,k}$ are defined by

$$F_{n,1}(x, y) = F_n(x, y),$$
$$F_{n,i}(x, y) = F_n[x, F_{n,i-1}(x, y)] \quad (i = 2, 3, \ldots, k). \tag{4.19}$$

The converse is generally false, as the following simple examples will show. Let $F_n(x, y) = F_{n,1}(x, y) = x + y^2$ be a function of the real variables x and y. Then $F_{n,2}(x, y) = x + (x + y^2)^2$. The equation $x = F_n(x, x)$, i.e., $x = x + x^2$, has the unique solution $x = 0$, whereas the equation $x = F_{n,2}(x, x)$, or $x = x + (x + x^2)^2$, has two solutions: $x = 0$ and $x = -1$. If $F_n(x, y) = x - y$, then $F_{n,k} = x - y$ for odd k and $F_{n,k}(x, y) = y$ for even k. Thus the equation $x = F_{n,k}(x, x)$ has a unique solution 0 for odd k but infinitely many solutions for even k.

However, if (4.1) has the same number of solutions on D as each of equations (4.18), $k, n = 1, 2, \ldots$, then equations (4.1) and (4.18) are equivalent on D. Consequently, solution of (4.1) reduces to solution of (4.18). The generalized iterative method is then defined by

$$x_n = F_{n,k}(x_n, x_{n-1}) \quad (n = 1, 2, \ldots). \tag{4.20}$$

We shall not, however, devote attention to algorithm (4.20) in the general, nonstationary case, preferring to present a detailed account of the stationary case, when the operators $F_{n,k}$ are independent of n.

§5. Stationary generalized iterative method for equations in a space R metrized by elements of a set G

In this section we consider the (not necessarily linear) operator equation

$$x = F(x, x), \tag{5.1}$$

where F depends on two variables, ranging over a domain in a space R metrized by elements of a set G (see §1), and takes values in R: $F(x, y) \in R$ for $x, y \in R$.

The gist of the method to be studied is that the successive approximations x_n to the desired solution of (5.1) are defined for each n as solutions of the operator equations

$$x_n = F_k(x_n, x_{n-1}) \quad (n = 1, 2, \ldots), \tag{5.2}$$

where the operators F_i $(i = 1, \ldots, k)$ are defined by the recurrence relations

$$\begin{gathered} F_1(x, y) = F(x, y), \\ F_i(x, y) = F[x, F_{i-1}(x, y)] \quad (i = 2, 3, \ldots, k); \end{gathered} \tag{5.3}$$

x_0 is an arbitrary element of R.

This algorithm was proposed by the author for operator equations in a Banach space in [197], and for a special case of the operator F in [100].

1. We shall assume that for each $n = 1, 2, \ldots$ equation (5.2) has a (not necessarily unique) solution x_n in a certain closed set $D \subset R$, $x_0 \in D$, and equation (5.1) has a solution x.

THEOREM 5.1. *Suppose that for all $x, y, z, t \in D$*

$$d(F(x, y), F(z, t)) \leqslant \varphi(d(x, z), d(y, t)), \tag{5.4}$$

where $\varphi(c, d)$ is a function of two variables, defined on the square $\Delta \times \Delta$, $\Delta = [\theta, k] \subset G$, with values in G, nondecreasing and jointly sequentially monotonically continuous, such that the equation $u = \varphi_k(u, u)$, where $\varphi_k(c, d)$ is defined by

$$\begin{gathered} \varphi_1(c, d) = \varphi(c, d), \\ \varphi_i(c, d) = \varphi[c, \varphi_{i-1}(c, d)] \quad (i = 2, 3, \ldots, k), \end{gathered} \tag{5.5}$$

has the unique solution $u = \theta$, and let $\varphi(b, b) \leqslant b$, where b is the diameter of D.

Then the sequence $\{x_n\}$, where x_n is a solution of equation (5.2) in D, converges to the unique solution x of equation (5.1) in D for any $x_0 \in D$, and the following error estimates are valid:

$$1)\ d(x, x_n) \leqslant B_{n,k}(b) \quad (n = 0, 1, 2, \ldots), \tag{5.6}$$

where $B_{n,k}(b)$ is an upper solution on $[\theta, b]$ of the equation

$$u = \varphi_k(u, B_{n-1,k}(b)), \quad B_{0,k}(b) = b; \tag{5.7}$$

$$2)\ d(x, x_n) \leqslant m(\psi_n, b), \tag{5.8}$$

where $m(\psi_n, b)$ is an upper solution of the equation

$$u = \psi_n(u) \equiv d(x_n, x_{n+1}) + \varphi_k[B_{n+1,k}(b), u]. \tag{5.9}$$

Moreover, $B_{n,k}(b) \searrow \theta$ as $n \to \infty$, for fixed k. If in addition

$$\varphi(u, u) \leqslant u \quad (u \in [\theta, b]), \tag{5.10}$$

then

$$B_{n,k}(b) \leqslant B_{n,l}(b) \quad \text{for} \quad k \geqslant l. \tag{5.11}$$

PROOF. We shall first establish a few auxiliary propositions, on which the main proof will be based.

We first show that for any k the function $\varphi_k(c, d)$ is jointly nondecreasing. Indeed, $\varphi_1(c, d) = \varphi(c, d)$ is a nondecreasing function by assumption. Suppose that $\varphi_i(c, d)$ is nondecreasing, i.e., $\varphi_i(c, d) \leqslant \varphi_i(g, h)$ for $c \leqslant g$ and $d \leqslant h$. Then

$$\varphi_{i+1}(c, d) = \varphi(c, \varphi_i(c, d)) \leqslant \varphi(g, \varphi_i(g, h)) = \varphi_{i+1}(g, h)$$

and so $\varphi_i(c, d)$ is jointly nondecreasing for all $i = 1, 2, \ldots$.

We now claim that for any $i = 2, 3, \ldots$,

$$\varphi_i(u, b) \leqslant \varphi_{i-1}(u, b) \quad (u \in [\theta, b]). \tag{5.12}$$

In view of the fact that $\varphi(b, b) \leqslant b$, we find that

$$\varphi_2(u, b) = \varphi(u, \varphi(u, b)) \leqslant \varphi(u, \varphi(b, b)) \leqslant \varphi(u, b) = \varphi_1(u, b).$$

If (5.12) is true for some i, then

$$\varphi_{i+1}(u, b) = \varphi(u, \varphi_i(u, b)) \leqslant \varphi(u, \varphi_{i-1}(u, b)) = \varphi_i(u, b),$$

so that it is true for any i.

It follows from (5.12) and the inequality $\varphi(b, b) \leqslant b$ that, in particular,

$$\varphi_i(b, b) \leqslant \varphi_{i-1}(b, b) \leqslant b \quad (i = 2, 3, \ldots). \tag{5.13}$$

We now prove that for any $x, y, z, t \in D$

$$d(F_k(x, y), F_k(z, t)) \leqslant \varphi_k(d(x, z), d(y, t)). \tag{5.14}$$

Indeed, this is obvious for $k = 1$. Supposing it true for $k = i$, we have

$$\begin{aligned} d(F_{i+1}(x, y), F_{i+1}(z, t)) &= d[F(x, F_i(x, y)), F(z, F_i(z, t))] \\ \leqslant \varphi[d(x, z), d(F_i(x, y), F_i(z, t))] &\leqslant \varphi[d(x, z), \varphi_i(d(x, z), d(y, t))] \\ &= \varphi_{i+1}(d(x, z), d(y, t)) \end{aligned}$$

and so by induction inequality (5.14) is true for any k.

We now prove the inequality

$$B_{n,k}(b) \leqslant B_{n-1,k}(b). \tag{5.15}$$

Since $\varphi_k(u, v)$ is nondecreasing, we deduce from (5.13) that

$$\varphi_k(u, B_{0,k}(b)) = \varphi_k(u, b) \leqslant \varphi_k(b, b) \leqslant b$$

Moreover, $\varphi_k(u, B_{0,k}(b))$ is a nondecreasing function of u and is easily seen to be sequentially monotonically continuous. Hence, by the functional inequality lemma (§1), equation (5.7) with $n = 1$ has an upper solution $B_{1,k}(b) \leqslant b$ on $[\theta, b]$. Thus $B_{1,k}(b) \leqslant B_{0,k}(b) = b$. Suppose now that (5.15) is true for some $k = i$. Then for any $u \in [\theta, b]$

$$\varphi_k(u, B_{i,k}(b)) \leqslant \varphi_k(u, B_{i-1,k}(b)), \tag{5.16}$$

and it follows by the same lemma that any solution of (5.7) with $n = i + 1$ cannot exceed the solution of that equation with $n = i$, i.e., $B_{i+1,k}(b) \leqslant B_{i,k}(b)$, and thus (5.15) is true for any $n = 1, 2, \ldots$ and $k = 1, 2, \ldots$.

Since $B_{0,k}(b) = b$, we also have

$$B_{n,k}(b) \leqslant b \tag{5.17}$$

for any n and k.

We now proceed to the proof of Theorem 5.1. Since by assumption all the solutions x_n of equations (5.2) lie in D, it follows that for all $m = 1, 2, \ldots$

$$d(x_m, x_0) \leqslant b = B_{0,k}(b).$$

Assuming that

$$d(x_{m+n}, x_n) \leqslant B_{n,k}(b) \tag{5.18}$$

for some value of n and all m, we find

$$\begin{aligned} d(x_{m+n+1}, x_{n+1}) &= d(F_k(x_{n+m+1}, x_{n+m}), F_k(x_{n+1}, x_n)) \\ &\leqslant \varphi_k(d(x_{n+m+1}, x_{n+1}), d(x_{n+m}, x_n)) \\ &\leqslant \varphi_k(d(x_{n+m+1}, x_{n+1}), B_{n,k}(b)). \end{aligned} \tag{5.19}$$

But since $d(x_{n+m+1}, x_{n+1}) \leqslant b$ (as all the x_k are in D), it follows, via the fact that $\varphi_k(c, d)$ is jointly nondecreasing and also by (5.17) and (5.13), that

$$\varphi_k(d(x_{n+m+1}, x_{n+1}), B_{n,k}(b)) \leqslant \varphi_k(b, b) \leqslant b.$$

Consequently, by the functional inequality lemma, it follows from inequality (5.19), with p in the lemma set equal to $d(x_{n+m+1}, x_{n+1})$, that

$$d(x_{n+m+1}, x_{n+1}) \leqslant B_{n+1,k}(b).$$

Thus by induction (5.18) is true for any n, m and k.

We claim that $B_{n,k} \searrow \theta$ for fixed k as $n \to \infty$. Indeed, the sequence

$\{B_{n,k}(b)\}$ of elements of G is nonincreasing and tends to some limit $c \leqslant b$. By definition

$$B_{n,k}(b) = \varphi_k(B_{n,k}(b), B_{n-1,k}(b)),$$

and so, letting $n \longrightarrow \infty$ and using the joint sequential monotonic continuity of $\varphi_k(c, d)$ (which follows from the analogous property of $\varphi(c, d)$), we obtain

$$c = \varphi_k(c, c), \tag{5.20}$$

whence it follows that $c = \theta$. Thus $B_{n,k}(b) \searrow \theta$ as $n \longrightarrow \infty$.

Now, noting inequality (5.18) and the fact that $B_{n,k}(b) \searrow \theta$ as $n \longrightarrow \infty$, using the completeness of the space R, we conclude that the sequence $\{x_n\}$ converges to some limit x. And since by assumption D is closed and $x_n \in D$, it follows that $x \in D$; thus, letting $n \longrightarrow \infty$ in (5.18), we obtain (5.6).

We now show that x satisfies the equation

$$x = F_k(x, x). \tag{5.21}$$

Indeed,

$$\begin{aligned} d(x, F_k(x, x)) &\leqslant d(x, x_n) + d(x_n, F_k(x, x)) \\ &\leqslant d(x, x_n) + d(F_k(x_n, x_{n-1}), F_k(x, x)) \\ &\leqslant d(x, x_n) + \varphi_k(d(x_n, x), d(x_{n-1}, x)) \\ &\leqslant B_{n,k}(b) + \varphi_k(B_{n,k}(b), B_{n-1,k}(b)) = B_{n,k}(b) + B_{n,k}(b), \end{aligned}$$

whence it follows that $d(x, F_k(x, x)) = \theta$, and x is a solution of (5.21).

It is readily seen that x is the unique solution of (5.21) on D. For let $\bar{x}$ be a solution in D other than x. Then, since $x_0 \in D$, we have $d(\bar{x}, x_0) \leqslant b = B_{0,k}(b)$. Supposing that

$$d(\bar{x}, x_n) \leqslant B_{n,k}(b) \leqslant b, \tag{5.22}$$

we see that

$$\begin{aligned} d(\bar{x}, x_{n+1}) &= d(F_k(\bar{x}, \bar{x}), F(x_{n+1}, x_n)) \\ &\leqslant \varphi_k(d(\bar{x}, x_{n+1}), d(\bar{x}, x_n)) \leqslant \varphi_k(d(\bar{x}, x_{n+1}), B_{n,k}(b)) \end{aligned}$$

and by the functional inequality lemma

$$d(\bar{x}, x_{n+1}) \leqslant B_{n+1,k}(b),$$

which means that (5.22) is true for all n and k. Then, as in §2, we obtain

$$d(x, \bar{x}) \leqslant d(x, x_n) + d(\bar{x}, x_n) \leqslant B_{n,k}(b) + B_{n,k}(b),$$

whence it follows that $d(x, \bar{x}) = \theta$ and $x = \bar{x}$.

Another consequence of the above argument is that all sequences $\{x_n\}$

($x_n \in D$) of solutions of equations (5.2) will converge to the same limit $x \in D$. In fact, taking any other sequence $\{x'_n\}$, where x'_n are solutions of (5.2), and reasoning as in the case of the sequence $\{x_n\}$, we show that it has some limit $x' \in D$ which is a solution of (5.21); but we have proved that this equation is uniquely solvable, and so $x' = x$.

The element x, which is the unique solution of equation (5.2) on D, is also the unique solution on D of (5.1), since as remarked in §4 each solution of (5.1) is a solution of (5.21), and by assumption (5.1) is solvable.

To establish (5.8), we reason as follows:

$$\begin{aligned} d(x_n, x) &\leqslant d(x_n, x_{n+1}) + d(x_{n+1}, x) \\ &\leqslant d(x_n, x_{n+1}) + d(F_k(x_{n+1}, x_n), F_k(x, x)) \\ &\leqslant d(x_n, x_{n+1}) + \varphi_k(d(x_{n+1}, x), d(x_n, x)) \\ &\leqslant d(x_n, x_{n+1}) + \varphi_k(B_{n+1,k}(b), d(x_n, x)). \end{aligned}$$

Hence, by the functional inequality lemma, we obtain (5.8).

It remains to prove (5.11). Since $B_{1,i}(b)$ is an upper solution of the equation $u = \varphi_i(u, b)$ on $[\theta, b]$ and $B_{1,i-1}(b)$ is an upper solution of the equation $u = \varphi_{i-1}(u, b)$ on $[\theta, b]$, it follows via (5.12) and the functional inequality lemma that $B_{1,i}(b) \leqslant B_{1,i-1}(b)$ $(i = 2, 3, \ldots)$. Next, assuming that the inequality

$$B_{n,i}(b) \leqslant B_{n,i-1}(b) \tag{5.23}$$

is true for some n and noting that by (5.10)

$$\varphi(B_{n,i-1}(b), B_{n,i-1}(b)) \leqslant B_{n,i-1}(b),$$

we find that

$$\varphi_i(u, B_{n,i}(b)) = \varphi_{i-1}(u, \varphi(u, B_{n,i}(b))) \leqslant \varphi_{i-1}(u, B_{n,i-1}(b))$$
$$(u \in [\theta, B_{n,i-1}(b)]),$$

and hence, by the lemma of §1,

$$B_{n+1,i}(b) \leqslant B_{n+1,i-1}(b).$$

Thus by induction inequality (5.23) is true for any n and i. Inequality (5.11) follows immediately from (5.23).

REMARK 1. Instead of assuming that the equation $u = \varphi_k(u, u)$ has a unique solution $u = \theta$, we may assume that the equation $u = \varphi(u, u)$ has a unique solution $u = \theta$ and that (5.10) holds.

To prove this, we use the inequality

$$\varphi_k(u, u) \leqslant \varphi_{k-1}(u, u), \tag{5.24}$$

which is easily verified by induction. It follows from this inequality that any upper solution of the equation $u = \varphi_k(u, u)$ on $[\theta, b]$ cannot exceed the upper solution of the equation $u = \varphi(u, u)$ on this interval. And since the latter equation has a unique solution there by assumption, the solution being θ, it follows that θ, as a solution of the equation $u = \varphi_k(u, u)$, is the unique solution of the equation on $[\theta, b]$.

REMARK 2. If the domain D is a ball $S(z, r)$ and the initial approximation x_0 is its center z, we may replace the diameter b of D throughout by the radius r of $S(z, r)$.

Indeed, in this case all the auxiliary propositions (5.12) through (5.15) clearly remain valid. Then, observing that

$$d(x_n, x_0) = d(x_n, z) \leqslant r = B_{0,k}(r),$$

we prove the inequality

$$d(x_{n+m}, x_n) \leqslant B_{n,k}(r)$$

in the same way as inequality (5.18), and so on.

2. In subsection 1 we proved inequality (5.23), which implies in particular that $\{B_{n,i}(b)\}$ $(i = 1, 2, \ldots)$, as a nonincreasing sequence of elements of G, converges to a limit c_n. We now look for conditions under which $c_n = \theta$. By (5.12) and the fact that $\theta \leqslant \varphi_i(u, b)$ for any $i = 1, 2, \ldots$, we deduce that $\varphi_i(u, b)$ is nonincreasing and converges to some limit $\Phi(u)$. We claim that if the equation

$$u = \Phi(u) \tag{5.25}$$

has no solution other than θ, then $c_n = \theta$, i.e., $B_{n,i} \searrow \theta$.

Indeed, since $B_{1,i}(b)$ solves the equation $u = \varphi_i(u, b)$, it follows that c_1 is a solution of (5.25), $c_1 = \Phi(c_1)$, and so $c_1 = \theta$. Now we have shown that

$$B_{n,i}(b) \leqslant B_{n-1,i}(b) \leqslant \ldots \leqslant B_{1,i}(b),$$

and so

$$\theta \leqslant c_n \leqslant c_{n-1} \leqslant \ldots \leqslant c_1 = \theta.$$

Thus $c_n = \theta$ for any n.

In the general case it is quite difficult to verify whether equation (5.25) has any nontrivial solutions, since Φ is usually not known. However, when the equation $u = \varphi(u, u)$ has no solutions other than θ, the same will be true of (5.25). This is proved by reductio ad absurdum. For any $u \in [\theta, b]$, we have

$$\varphi_k(u, b) = \varphi(u, \varphi_{k-1}(u, b)) \tag{5.26}$$

and, since φ is sequentially monotonically continuous,

$$\Phi(u) = \varphi(u, \Phi(u)) \tag{5.27}$$

for any $u \in [\theta, b]$. Supposing that (5.25) has a nontrivial solution c, $c = \Phi(c)$, and setting $u = c$ in (5.27), we obtain $c = \varphi(c, c)$, contrary to the assumption that the equation $u = \varphi(u, u)$ has no solutions other than θ.

3. Our proof of Theorem 5.1 was based on the assumption that equations (5.2) have solutions x_n lying in some set D. It is not hard to derive conditions guaranteeing this situation. If the operator $F(x, y)$ maps $D \times D$ into D, i.e., $F(x, y) \in D$ for all $x, y \in D$, then it is readily seen that for all $k = 1, 2, \ldots$ the operator $F_k(x, y)$ also maps $D \times D$ into D. Therefore, if $F(x, y)$, as a function of x satisfies the inequality

$$d(F_k(x, t),\ F_k(y, t)) \leqslant a_k(d(x, y)), \tag{5.28}$$

where the functions $a_k(c)$ have the same properties as the function $A(c)$ in Theorem 2.2, then equation (5.2) has a unique solution x_n in D for each n, for any initial approximation $x_0 \in D$.

We shall show that $a_k(c)$ will have the same properties as $A(c)$ if the function $\varphi(c, d)$ defined by (5.4) is nondecreasing and sequentially monotonically continuous in each of its arguments, the equation $u = \varphi(u, u)$ has a unique solution $u = \theta$, and inequality (5.10) holds.

In fact, we have

$$a_k(c) = \varphi_k(c, \theta). \tag{5.29}$$

That $a_k(c)$ is nondecreasing and sequentially monotonically continuous is obvious. It is also obvious that

$$a_k(b) \leqslant b. \tag{5.30}$$

We shall show that the equation $u = a_k(u)$ has a unique solution $u = \theta$. Clearly, θ is a solution of this equation, since it is a solution of the equation $u = \varphi(u, u)$. Next, since

$$\varphi(u, \theta) \leqslant \varphi(u, u) \quad (u \in [\theta, b]),$$

it follows by the functional inequality lemma (§1) that the equation $u = \varphi(u, \theta)$ has an upper solution which does not exceed the upper solution of the equation $u = \varphi(u, u)$ and is therefore equal to θ. If c is some other montrivial solution of the equation $u = a_k(u)$, then

$$c = a_k(c) = \varphi_k(c, \theta) = \varphi(c, \varphi_{k-1}(c, \theta)) \leqslant \varphi(c, c).$$

Hence, since by assumption inequality (5.10) is true, it follows that $c = \theta$.

In the case $k = 1$, in order to prove that $a_1(c)$ has the properties of $A(c)$ of

Theorem 1.2, it is clearly sufficient to require that $\varphi(c, d)$ be nondecreasing and sequentially monotonically continuous only in the first argument, the equation $u = \varphi(u, \theta)$ has a unique solution θ, and $\varphi(b, \theta) \leqslant b$.

If the set D is taken to be a ball $S(z, r)$, and $\varphi(u, v)$ is nondecreasing in each variable, we can ensure that $F_k(x, y)$ maps $S(z, r) \times S(z, r)$ into $S(z, r)$ by demanding only that

$$q + \varphi(r, r) \leqslant r, \tag{5.31}$$

where $d(Tz, z) \leqslant q$.

Indeed, in this case we have, for any $x, y \in S(z, r)$,

$$\begin{aligned} d(F(x, y), z) &\leqslant d(F(x, y), F(z, z)) + d(F(z, z), z) \\ &\leqslant q + \varphi(d(x, z), d(y, z)) \leqslant q + \varphi(r, r) \leqslant r, \end{aligned}$$

so that $F(x, y)$ maps $S(z, r) \times S(z, r)$ into $S(z, r)$. Thus the operator $F_k(x, y)$ (k arbitrary) also maps $S(z, r)$ into $S(z, r)$.*

We have thus proved the following theorem.

THEOREM 5.2. *Suppose that inequality* (5.4) *is true for arbitrary* $x, y, z, t \in S(z, r)$, *with* $\varphi(c, d)$ *a function defined on* $[0, b] \times [0, b]$, *nondecreasing and sequentially monotonically continuous in both arguments, such that the equation* $u = \varphi(u, u)$ *has a unique solution* θ, $\varphi(b, b) \leqslant b$ *for* $2r \leqslant b$, *and inequality* (5.31) *holds.*

Then equation (5.2) *is uniquely solvable in the ball* $S(z, r)$ *for each* $n = 1, 2, \ldots$ *and any initial approximation* $x_0 \in S(z, r)$.

Moreover, if $x_0 = z$, b *may be replaced by* r *throughout.*

4. We now consider the question of determining the radius of a ball which remains invariant under the operator $F(x, y)$. Let $S(z, r)$ be a ball whose center z is a fixed point of the space R; r will be determined below. Suppose that for all $x, y \in S(z, r)$

$$d(F(x, y), z) \leqslant V(r), \tag{5.32}$$

where $V(r)$ is a known function defined on G and with values in G. If there exists $r = r^*$ such that

$$V(r^*) \leqslant r^*, \tag{5.33}$$

then $F(x, y)$ maps the ball $S(z, r^*)$ into itself: $F(x, y) \in S(z, r^*)$ for $x, y \in S(z, r^*)$.

---

* ***Translator's note.*** **The author employs this loose terminology frequently in analogous situations (the operator in question of course actually maps $S(z, r) \times S(z, r)$ into $S(z, r)$).**

The proof is obvious.

The radius of a ball invariant under $F(x, y)$ may also be found as follows. Suppose that for all $x, y \in S(z, r)$

$$d(F(x, y), F(z, z)) \leqslant W(r), \tag{5.34}$$

where $W(r)$ is a function defined on G with values in G. Then the required radius is a value of r^* satisfying the inequality

$$W(r^*) + d(F(z, z), z) \leqslant r^*. \tag{5.35}$$

Indeed, if this is true, then for any $x, y \in S(z, r^*)$ we have

$$\begin{aligned} d(F(x, y), z) &\leqslant d(F(x, y), F(z, z)) + d(F(z, z), z) \\ &\leqslant W(r^*) + d(F(z, z), z) \leqslant r^*, \end{aligned}$$

so that $F(x, y)$ maps $S(z, r^*)$ into itself.

It is often desirable to select the smallest of all values r^* satisfying inequality (5.33) or (5.35), if it exists. Conditions for the existence of such numbers r^* may be derived using the functional inequality lemma of §1.

5. We now compare the generalized iterative method defined by (5.2) and (5.3) with the usual method of successive approximations, according to which the approximations y_n to the solution of (5.1) are determined by

$$y_n = F(y_{n-1}, y_{n-1}). \tag{5.36}$$

Suppose that the operator $F(x, y)$ maps a closed subset D of R into itself and condition (5.4) holds for arbitrary $x, y, z, t \in D$, where the function $\varphi(c, d)$, defined on the set $\Delta \times \Delta$ $(\Delta = [\theta, k], k \in G)$, is jointly nondecreasing, sequentially monotonically continuous, and the equation $u = \varphi(u, u)$ has the unique solution $u = \theta$.

Then there clearly exists a function $A(c)$ such that, for any $x, y \in D$,

$$\begin{aligned} d(F(x, x), F(y, y)) &\leqslant A(d(x, y)), \\ A(c) &\leqslant \varphi(c, c) \quad (c \in [\theta, b]), \end{aligned} \tag{5.37}$$

and so the successive approximations (5.36) converge to a solution of (5.1); the error estimate for the nth approximation is

$$d(x, y_n) \leqslant A_n(b), \tag{5.38}$$

where $A_n(b)$ is defined by the recurrence relations

$$A_0(b) = b, \quad A_n(b) = A(A_{n-1}(b)) \quad (n = 1, 2, \ldots) \tag{5.39}$$

(see §2).

Under these assumptions, the algorithm (5.2) also converges when $k = 1$, and if in addition $\varphi(u, u) \leqslant u$ for all $u \in [\theta, b]$, it converges for any k. However, the processes (5.2) and (5.36) may well converge at different rates. Some idea of the difference may be gained from the above estimates, though they are extremely "generous" and therefore do not reflect the real situation: if $A_n(b) \leqslant B_{n,k}(b)$, $b = 0, 1, \ldots$, the algorithm (5.36) is more rapid, while if $B_{n,k}(b) \leqslant A_n(b)$, $b = 0, 1, \ldots$, the faster algorithm is (5.2).

We claim that if $A(c) = \varphi(c, c)$, then

$$B_{n,1}(b) \leqslant A_n(b). \tag{5.40}$$

Indeed, since in this case

$$A_1(b) = A(A_0(b)) = A(b) = \varphi(b, b),$$

and $B_{1,1}(b)$ is an upper solution of the equation $u = \varphi(u, b)$ on $[\theta, b]$, it follows that $B_{1,1}(b) \leqslant A_1(b)$. Supposing (5.40) true for some n, we get

$$B_{n+1,1}(b) = \varphi(B_{n+1,1}(b),\ B_{n,1}(b)) \leqslant \varphi(B_{n,1}(b),\ B_{n,1}(b))$$
$$\leqslant \varphi(A_n(b),\ A_n(b)) = A_{n+1}(b),$$

so that (5.40) is true for every n.

If $B_{n,k}(b) \leqslant B_{n,k-1}(b)$, $k = 2, 3, \ldots$, then it is clear that

$$B_{n,k}(b) = A_n(b), \tag{5.41}$$

so that the error estimate (5.6) is sharper than (3.7).

Note that the convergence domains of algorithms (5.2) and (5.36) are generally different. In other words, there are equations for which the first algorithm converges but the second diverges, and vice versa.

Moreover, it may happen that the same algorithm (5.2) will converge for one value of k of a solution of (5.1), but diverge for another value. However, if the operator $F(x, y)$ is suitably chosen, algorithm (5.2) will usually have a larger convergence domain. Various special cases of the operator $F(x, y)$, as well as conditions for convergence of the successive approximations defined by (5.2) and error estimates, will be considered in Chapter II for operator equations in Banach spaces, and in Chapter III for some special types of operator equations.

To conclude this section, we note that if there exists an operator Φ_k (the resolvent operator) such that for any $y \in D$ the solution of the equation

$$x = F_k(x, y) \tag{5.42}$$

is defined by

$$x = \Phi_k(y), \tag{5.43}$$

then the successive approximation process (5.2) is equivalent to the usual successive approximation procedure applied to the equation

$$x = \Phi_k(x), \tag{5.44}$$

i.e., to the algorithm

$$x_n = \Phi_k(x_{n-1}) \quad (n = 1, 2, \ldots). \tag{5.45}$$

However, since in general no expression is available for the operator Φ_k, convergence conditions involving Φ_k are frequently rather difficult to check, and are therefore not too effective. This is also true of the corresponding error estimates.

§6. Nonstationary generalized iterative method for equations in lattice-normed spaces

In this section we study a generalized iterative method in which the approximations x_n to the solution of the equation

$$x = Tx \tag{6.1}$$

in a space E normed by the elements of an archimedean lineal N are defined as solutions of the equations

$$x_n = T_n(x_n, x_{n-1}) \quad (n = 1, 2, \ldots), \tag{6.2}$$

where T_n are operators from E to E and x_0 is an element of E.

Suppose that for all solutions x of (6.1) in some closed subset D of E

$$\lim \| T_n(x, x) - Tx \| = 0 \tag{6.3}$$

in the sense of uniform convergence (see §1).

THEOREM 6.1. *Suppose that equation* (6.1) *has a solution x in D, and for each $n = 1, 2, \ldots$ the operator equation* (6.2) *has a solution x_n in the same set. Let the operators T_n have the property that for any $x, y, z, t \in D$*

$$\| T_n(x, y) - T_n(z, t) \| \leqslant M_n \| x - z \| + K_n \| y - t \|, \tag{6.4}$$

where, for any operators M and P such that

$$M_n \leqslant M, \ K_n \leqslant K, \ P = (I - M)^{-1} K,$$

the series $\Sigma_0^\infty M^i a$ and $\Sigma_0^\infty P^i a$ converge uniformly for any $a \in N$.

Then equation (6.1) *has a unique solution in D, and* (6.3) *is a necessary and sufficient condition for convergence of a sequence $\{x_n\}(x_n \in D)$ of solutions of equations* (6.2) *to the solution of* (6.1) *for any initial approximation $x_0 \in D$. The following error estimates hold for the nth approximation*:

$$\text{a)} \quad \| x - x_n \| \leqslant (I - M_n)^{-1} K_n \| x - x_{n-1} \| + (I - M_n)^{-1} a_n \tag{6.5}$$

$$(a_n = \| T_n(x, x) - Tx \|);$$

(b) $$\|x - x_n\| \leqslant P^n \|x - x_0\| + \sum_{i=0}^{n-1} P^i (I - M)^{-1} a_{n-i}. \tag{6.6}$$

PROOF. To prove that condition (6.3) is sufficient, we establish the estimate

$$\begin{aligned} \|x_n - x\| &= \|T_n(x_n, x_{n-1}) - Tx\| \leqslant \|T_n(x_n, x_{n-1}) - T_n(x, x)\| \\ &+ \|T_n(x, x) - Tx\| \leqslant M_n \|x_n - x\| + K_n \|x_{n-1} - x\| + \|T_n(x, x) - Tx\| \\ &\leqslant M\|x_n - x\| + K\|x_{n-1} - x\| + \|T_n(x, x) - Tx\|. \end{aligned} \tag{6.7}$$

Since $\Sigma_0^\infty M^i a$ is uniformly convergent, this inequality implies

$$\|x_n - x\| \leqslant P\|x_{n-1} - x\| + b_n \quad (b_n = (I - M)^{-1} a_n). \tag{6.8}$$

The rest of the proof that $\{x_n\}$ converges to x is entirely analogous to the proof in [168]. Since by assumption the sequence $\{a_n\}$ converges uniformly to θ and the operator $(I - M)^{-1}$ is continuous, the sequence $\{b_n\}$ also converges uniformly to θ. Let n' denote the largest natural number not exceeding $n/2$. Then, by (6.8),

$$\|x_n - x\| \leqslant P^n \|x_0 - x\| + \sum_{i=0}^{n'} P^i b_{n-i} + \sum_{i=n'+1}^{n-1} P^i b_{n-i}. \tag{6.9}$$

We claim that each of the three terms on the right of this inequality converges uniformly to θ. We have

$$\begin{aligned} \lim_{n\to\infty} P^n \|x_0 - x\| &= \lim_{n\to\infty} \left(\sum_{i=0}^{n} P^i \|x_0 - x\| - \sum_{i=0}^{n-1} P^i \|x_0 - x\| \right) \\ &= \lim_{n\to\infty} \sum_{i=0}^{n} P^i \|x_0 - x\| - \lim_{n\to\infty} \sum_{i=0}^{n-1} P^i \|x_0 - x\| = \theta. \end{aligned}$$

Next, since $\{b_n\}$ converges uniformly to θ, there exist $b \in N$ and a monotonically decreasing sequence of numbers $\{\lambda_n\}$ $(\lambda_n \leqslant \lambda)$, converging to θ, such that $b_n = |b_n| \leqslant \lambda_n b \leqslant \lambda b$, $n = 1, 2, \ldots$. Therefore,

$$\theta \leqslant \sum_{i=n'+1}^{n} P^i b_{n-i} \leqslant \lambda \sum_{i=n'+1}^{n} P^i b = \lambda \sum_{i=0}^{n} P^i b - \lambda \sum_{i=0}^{n'} P^i b,$$

and since $\Sigma_1^\infty P^i b$ is uniformly convergent for any $b \in N$ it follows that the third term in (6.9) converges uniformly to θ. Finally, since $b_i \leqslant \lambda_i b \leqslant \lambda_{n'} b$ for $i \geqslant n'$, it follows that

$$\theta \leqslant \sum_{i=0}^{n'} P^i b_{n-i} \leqslant \lambda_{n'} \sum_{i=0}^{n} P^i b \leqslant \lambda_{n'} \sum_{i=0}^{\infty} P^i b = \lambda_{n'} c \quad (c \in N)$$

and since $\lambda_{n'} \longrightarrow 0$ as $n' \longrightarrow \infty$ the second term on the right of (6.9) converges uniformly to θ. This completes the sufficiency proof.

Under the assumptions of Theorem 6.1, equation (6.1) has a unique solution.

Indeed, were x' a solution of equation (6.1) in D, distinct from x, we could prove as before that the sequence $\{x_n\}$ converges to this solution, and so the same sequence would converge to two distinct limits, which is impossible.

Note that by condition (6.4), for each n, equation (6.2) is uniquely solvable for x_n in the set D. In fact, the existence of two solutions $x_n \in D$ and $x_n' \in D$ would imply that

$$\|x_n - x_n'\| = \|T_n(x_n, x_{n-1}) - T_n(x_n', x_{n-1})\| \leqslant M_n\|x_n - x_n'\| \leqslant M\|x_n - x_n'\|$$

and, repeating the argument, we would have

$$\|x_n - x_n'\| \leqslant M^i\|x_n - x_n'\|,$$

Hence, letting $i \longrightarrow \infty$, we see that $x_n = x_n'$.

The error estimates (6.5) and (6.6) follow in an obvious manner from (6.7) and (6.8).

That (6.3) is a necessary condition for convergence of $\{x_n\}$ to a solution x of equation (6.1) follow from the inequality

$$\|T_n(x, x) - Tx\| \leqslant \|T_n(x, x) - T_n(x_n, x_{n-1})\| + \|x_n - x\| \leqslant M\|x - x_n\| + K\|x - x_{n-1}\| + \|x_n - x\|,$$

if we let $n \longrightarrow \infty$.

REMARK 1. If the equation

$$x = T_n(x, x) \tag{6.10}$$

is equivalent on D to equation (6.1) for each n, then $a_n = \theta$ and the error estimates (6.5) and (6.6) are simplified.

2. Suppose that equation (6.2) has a solution x_n for each $n = 1, 2, \ldots$ in a closed subset D of E, and that

$$\lim\|T_n(x_{n-1}, x_{n-1}) - Tx_{n-1}\| = \theta, \tag{6.11}$$

in the sense of uniform convergence.

THEOREM 6.2 [168]. *Let equation* (6.1) *have a solution x in D, and suppose that for arbitrary x,* $y \in D$

$$\|T_n(x, x_{n-1}) - T_n(y, x_{n-1})\| \leqslant M\|x - y\|, \tag{6.12}$$

$$\|[Tx - T_n(x, x_{n-1})] - [Ty - T_n(y, x_{n-1})]\| \leqslant L\|x - y\|, \tag{6.13}$$

and the series $\Sigma_0^\infty M^i a$ *and* $\Sigma_0^\infty P^i a$, *where* $P = (I - M)^{-1}$, *are uniformly convergent for any* $a \in N$.

Then equation (6.1) *is uniquely solvable on D, and* (6.11) *is a necessary and sufficient condition for the sequence* $\{x_n\}$ ($x_n \in D$ *is a solution of equation* (6.2)) *to converge to the solution x of* (6.1) *for any initial approximation* $x_0 \in$

D. The following error estimates hold for the nth approximation:

a) $\|x - x_n\| \leqslant P\|x - x_{n-1}\| + c_{n-1};$ (6.14)

b) $\|x - x_n\| \leqslant P^n\|x - x_0\| + \sum_{i=0}^{n-1} P^i c_{n-i-1};$ (6.15)

c) $\|x - x_n\| \leqslant (I - P)^{-1}P\|x_n - x_{n-1}\| + (I - P)^{-1}c_{n-1},$ (6.16)

where

$$c_{n-1} = (I - M)^{-1}\|T_n(x_{n-1}, x_{n-1}) - Tx_{n-1}\|.$$

PROOF. It follows from the equalities $x_n = T_n(x_n, x_{n-1})$ and $x = Tx$ that

$$\begin{aligned}\|x_n - x\| &= \|T_n(x_n, x_{n-1}) - Tx\| \leqslant \|T_n(x_n, x_{n-1}) - T_n(x, x_{n-1})\| \\ &+ \|[Tx_{n-1} - T_n(x_{n-1}, x_{n-1})] - [Tx - T_n(x, x_{n-1})]\| \\ &+ \|T_n(x_{n-1}, x_{n-1}) - Tx_{n-1}\| \leqslant M\|x_n - x\| \\ &+ L\|x_{n-1} - x\| + \|T_n(x_{n-1}, x_{n-1}) - Tx_{n-1}\|,\end{aligned}$$

and so, since the operator $(I - M)^{-1}$ exists and is continuous (as follows from the uniform convergence of $\Sigma_0^\infty M^i a$), we deduce the estimates (6.14) and (6.15). The error estimate (6.16) follows from the inequality

$$\|x - x_n\| \leqslant P\|x - x_n\| + P\|x_n - x_{n-1}\| + c_{n-1},$$

which is derived in an obvious manner from (6.14).

Since it follows from the assumptions of the theorem that the sequence $\{c_n\}$ converges uniformly to θ, the proof that (6.11) is a sufficient condition for $\{x_n\}$ to converge to a solution of (6.1) is analogous to the proof carried out in subsection 1 for conditon (6.3).

The necessity of condition (6.11) follows directly from the inequality

$$\begin{aligned}\|T_n(x_{n-1}, x_{n-1}) - Tx_{n-1}\| &\leqslant \|T_n(x_{n-1}, x_{n-1}) - T_n(x_n, x_{n-1})\| \\ &+ \|T_n(x_n, x_{n-1}) - Tx\| + \|[Tx - T_{n-1}(x, x_{n-2})] \\ &- [Tx_{n-1} - T_{n-1}(x_{n-1}, x_{n-2})]\| + \|T_{n-1}(x, x_{n-2}) - T_{n-1}(x_{n-1}, x_{n-2})\| \\ &\leqslant M\|x_n - x_{n-1}\| + \|x_n - x\| + L\|x_{n-1} - x\| + M\|x_{n-1} - x\|,\end{aligned}$$

by letting $n \to \infty$.

Finally, the uniqueness of the solution of (6.1) on D is proved as in subsection 1.

3. We now consider the case that

$$T_n(x_{n-1}, x_{n-1}) = Tx_{n-1} \quad (n = 1, 2, \ldots), \tag{6.17}$$

but the existence of a solution to equation (6.1) on the set D is not assumed. We have the following existence and uniqueness theorem.

THEOREM 6.3. *Suppose that the operators T and T_n map a closed subset D of E into itself: $Tx \in D$ and $T_n(x, y) \in D$ for $x, y \in D$; let conditions* (6.12) *and* (6.13) *hold for all $x, y \in D$, and assume that the series $\Sigma_0^\infty M^i a$ and $\Sigma_0^\infty P^i a$, where $P = (I - M)^{-1}L$, are uniformly convergent for any $a \in N$. Finally, let condition* (6.17) *be satisfied.*

Then equation (6.2) *is uniquely solvable for x_n in D, for any n and $x_0 \in D$, and the sequence $\{x_n\}$ converges to a unique solution of equation* (6.1) *in D for any $x_0 \in D$. The following error estimate holds for the nth approximation:*

$$\|x - x_n\| \leqslant (I - P)^{-1} P^{n-p+1} \|x_p - x_{p-1}\| \quad (1 \leqslant p \leqslant n+1). \tag{6.18}$$

PROOF. The unique solvability of equations (6.2), $n = 1, 2, \ldots$, follows directly from the assumptions of the theorem by Theorem 2.2.

It follows from (6.2) that

$$\begin{aligned}\|x_n - x_{n-1}\| &= \|T_n(x_n, x_{n-1}) - T_{n-1}(x_{n-1}, x_{n-2})\| \\ &\leqslant \|T_n(x_n, x_{n-1}) - T_n(x_{n-1}, x_{n-1})\| + \|T_n(x_{n-1}, x_{n-1}) - Tx_{n-1}\| \\ &+ \|[Tx_{n-1} - T_{n-1}(x_{n-1}, x_{n-2})] - [Tx_{n-2} - T_{n-1}(x_{n-2}, x_{n-2})]\| \\ &+ \|Tx_{n-2} - T_{n-1}(x_{n-2}, x_{n-2})\| \leqslant M\|x_n - x_{n-1}\| + L\|x_{n-1} - x_{n-2}\|,\end{aligned}$$

whence

$$\|x_n - x_{n-1}\| \leqslant P\|x_{n-1} - x_{n-2}\| \leqslant P^{n-p}\|x_p - x_{p-1}\| \quad (1 \leqslant p \leqslant n-1). \tag{6.19}$$

Next, we have

$$\begin{aligned}\|x_{n+m} - x_n\| &\leqslant \|x_{n+m} - x_{n+m-1}\| + \ldots + \|x_{n+1} - x_n\| \\ &\leqslant (P^{m-1} + \ldots + P + I) P^{n-p+1} \|x_p - x_{p-1}\| \\ &\leqslant \sum_{i=0}^{\infty} P^i P^{n-p+1} \|x_p - x_{p-1}\| = (I - P)^{-1} P^{n-p+1} \|x_p - x_{p-1}\| \\ &\qquad (1 \leqslant p \leqslant n+1).\end{aligned} \tag{6.20}$$

Since $\lim P^n \|x_1 - x_0\| = \theta$, it follows that $\{x_n\}$ is a Cauchy sequence and so, since D is closed, has a limit $x \in D$.

The estimate (6.18) is obtained from (6.20) by letting $m \to \infty$.

We now show that the limit x satisfies (6.1). Indeed,

$$\begin{aligned}\theta \leqslant \|Tx - x\| &\leqslant \|Tx - T_n(x_n, x_{n-1})\| + \|x_n - x\| \\ &\leqslant \|[Tx - T_n(x, x_{n-1})] - [Tx_{n-1} - T_n(x_{n-1}, x_{n-1})]\| \\ &+ \|T_n(x, x_{n-1}) - T_n(x_n, x_{n-1})\| + \|Tx_{n-1} - T_n(x_{n-1}, x_{n-1})\| \\ &+ \|x_n - x\| \leqslant L\|x - x_{n-1}\| + M\|x - x_n\| + \|x_n - x\|,\end{aligned}$$

whence it follows that $Tx = x$.

The uniqueness of the solution is readily proved indirectly. The existence of another solution $x' \in D$ of (6.1) would imply as before that the sequence $\{x_n\}$ also converges to x', yielding a contradiction.

REMARK 2. If the equation

$$x = T_n(x, y) \tag{6.21}$$

has a solution x in D for all $y \in D$, and

$$x = \Phi_n y, \tag{6.22}$$

where Φ_n is some operator, then the successive approximation process (6.2) reduces to the usual nonstationary iterative process for the equation $x = \Phi_n(x)$ considered in §3.

REMARK 3. If the operators $T_n(x, y)$ can be expressed as

$$T_n(x, y) = R_n x + S_n y,$$

where R_n and S_n are operators, with R_n linear operators whose spectra do not contain 1, then $(I - R_n)^{-1}$ exists and we may express Φ_n as

$$\Phi_n x = (I - R_n)^{-1} S_n x.$$

REMARK 4. If the operator $T_n(x, y)$ is singular with respect to the first argument, i.e.,

$$T_n(x, y) = \sum_{i=1}^{k} \Phi_{in}(x, y)\varphi_i + \Psi_n y,$$

where the φ_i are elements of E, the Φ_{in} are functionals, the Ψ_n operators and k a finite number, then the solution of equations (6.2) at each step reduces to solution of a finite system of algebraic or transcendental equations and the evaluation of certain functionals.

§7. Some special cases of the generalized iterative method

The generalized iterative method for solution of the operator equation (6.1), as defined by (6.2), includes many different approximate methods, both iterative and iterative projection. We indicate a few of them.

1. The usual method of successive approximations for solving the operator equation

$$x = Tx \tag{7.1}$$

in a space E, which is defined by the recurrence relation

$$x_n = Tx_{n-1} \quad (n = 1, 2, \ldots), \tag{7.2}$$

is obtained from (6.2) by setting $T_n(x, y) = Ty$. But if $T_n(x, y) = T_n y$, where the T_n are operators from E to E, we have the classical nonstationary method of successive approximations

$$x_n = T_n x_{n-1} \quad (n = 1, 2, \ldots). \tag{7.3}$$

2. Let the elements of the space E be vectors x with components $x_1, \ldots, x_k$, where the x_i are elements of spaces E_i, and let the operator T be defined by

$$Tx = \{T_i(x_1, x_2, \ldots, x_k)\} \quad (i = 1, 2, \ldots, k),$$

where the T_i are operators from E to E_i.

Then, setting

$$T_n(x, y) = \{T_i(x_1, x_2, \ldots, x_{i-1}, y_i, y_{i+1}, \ldots, y_k)\},$$

we obtain the Seidel method:

$$x_{i,n} = T_i(x_{1,n}, \ldots, x_{i-1,n}, x_{i,n-1}, \ldots, x_{k,n-1}) \\ (i = 1, 2, \ldots, k). \tag{7.4}$$

If

$$T_n(x, y) = \{T_{i,n}(x_1, \ldots, x_{i-1}, y_i, \ldots, y_k)\},$$

where the $T_{i,n}$ are operators from E to E_i, we have the nonstationary Seidel method:

$$x_{i,n} = T_{i,n}(x_{1,n}, \ldots, x_{i-1,n}, x_{i,n-1}, \ldots, x_{k,n-1}) \\ (i = 1, 2, \ldots, k). \tag{7.5}$$

In the case that

$$T_i(x_1, x_2, \ldots, x_k) = \sum_{j=1}^{k} a_{ij}x_j + b_i,$$

where a_{ij}, x_j and b_i are real numbers, we obtain the classical Seidel method for systems of linear algebraic equations.

3. The Newton-Kantorovič method for equation (7.1) may be written as

$$x_n = Tx_{n-1} + T'(x_{n-1})(x_n - x_{n-1}) \quad (n = 1, 2, \ldots), \tag{7.6}$$

where $T'(x_{n-1})$ is the derivative of T at x_{n-1}. This method is deduced from (6.2) by setting

$$T_n(x, y) = Ty + T'(y)(x - y).$$

If the successive approximations are defined by

$$x_n = Tx_{n-1} + T'(x_0)(x_n - x_{n-1}), \tag{7.7}$$

we have modified Newton-Kantorovič method:

$$T_n(x, y) = Ty + T'(x_0)(x - y).$$

4. A more general iterative method

$$x_n = Tx_{n-1} + S_n(x_n - x_{n-1}) \tag{7.8}$$

which includes the basic and modified Newton-Kantorivoč methods as special cases, is obtained from (6.2) by setting

$$T_n(x, y) = Ty + S_n(x - y),$$

where the S_n are linear operators from E to E. In particular, if $S_n = \alpha_n I$, where the α_n are numbers and I the identity operator, we have the algorithm defined by

$$x_n = Tx_{n-1} + \alpha_n(x_n - x_{n-1}). \tag{7.9}$$

5. Various iterative methods of higher orders are also special cases of the generalized iterative method (6.2).

6. The algorithms (4.20) and (5.2) are also special cases of (6.2). For example, (5.2) is obtained by setting $T_n(x, y) = F_k(x, y)$, with the operators F_k defined by

$$\begin{gathered} F_1(x, y) = F(x, y), \\ F_i(x, y) = F(x, F_{i-1}(x, y)) \quad (i = 2, 3, \ldots, k) \end{gathered} \tag{7.10}$$

(assuming that $F(x, x) = Tx$).

In the case $F(x, y) = Rx + Sy$, where R and S are operators from E to E and S is linear, then $F_k(x, y)$ may be expressed as

$$F_k(x, y) = \sum_{i=0}^{k-1} S^i Rx + S^k y$$

and the algorithm (5.2) is defined by

$$x_n = \sum_{i=0}^{k-1} S^i Rx_n + S^k x_{n-1}. \tag{7.11}$$

7. The scheme of the generalized iterative method (6.2) also encompasses iterative projection methods, which generalize Sokolov's method of averaged functional corrections. We illustrate the idea of these methods as applied to equation (7.1).

Let P_n and P'_n be linear projections, i.e., $P_n^2 = P_n$ and $P'^2_n = P'_n$, projecting the space E onto subspaces E_{P_n} and $E_{P'_n}$, respectively. Setting

$$T_n(x, y) = P_nTx + Q_nT(P'_nx + Q'_ny),$$

where $Q_n = I - P_n$ and $Q'_n = I - P'_n$ (I is the identity operator), we obtain a non-stationary iterative projection method in which the successive approximations x to the solution of (7.1) are defined as solutions of the equations

$$x_n = P_nTx_n + Q_nT(P'_nx_n + Q'_nx_{n-1}) \qquad (n = 1, 2, \ldots, \quad x_0 \in E). \tag{7.12}$$

If $P_n = P$ and $P'_n = P'$, then we have the stationary iterative projection method

$$x_n = PTx_n + QT(P'x_n + Q'x_{n-1}) \qquad (n = 1, 2, \ldots, x_0 \in E), \tag{7.13}$$

where $Q = I - P$ and $Q' = I - P'$.

As special cases of (7.13), we have the algorithms

a) $x_n = PTx_n + QTx_{n-1};$ (7.14)

b) $x_n = T(P'x_n + Q'x_{n-1});$ (7.15)

c) $x_n = PTx_n + QT(Px_n + Qx_{n-1}),$ (7.16)

which are obtained by setting $P' = 0$, $P = 0$ and $P' = P$, respectively.

8. We can also construct an algorithm of type (5.2) with the operator F defined by

$$F(x, y) = PTx + QT(P'x + Q'y).$$

If T can be expressed as $Tx = f + Ax$, where $f \in E$ and A is a linear operator, the algorithm may be written

$$x_n = \sum_{i=0}^{k-1} (QAQ')^i f + \sum_{i=0}^{k-1} (QAQ')^i (PA + QAP')x_n + (QAQ')^k x_{n-1}. \tag{7.17}$$

Setting $P' = 0$, $P = 0$ and $P' = P$, respectively, we obtain the following algorithms:

$$x_n = \left[I + \sum_{i=0}^{k-2} (QAQ)^i QA\right](f + PAx_n) + (QAQ)^{k-1}QAx_{n-1}, \tag{7.18}$$

$$x_n = \left[I + AQ' \sum_{i=0}^{k-2} (Q'AQ')^i\right](f + AP'x_n) + AQ'(Q'AQ')^{k-1}Qx_{n-1}, \tag{7.19}$$

and

$$x_n = \sum_{i=0}^{k-1} (QAQ)^i [f + (PA + QAP)]x_n + (QAQ)^k Qx_{n-1}. \tag{7.20}$$

Convergence criteria and error estimates for the above methods may be derived directly from the general results. However, the convergence criteria and error estimates thus obtained may be too coarse, since they do not take the specific features of each method into consideration. For this reason, the fact that we have a general framework for all these methods does not eliminate the need for an individual treatment, aimed at establishing more accurate criteria for their application and sharper error estimates which allow for the specific properties of each method.

§8. Investigation of the stationary generalized iterative method for equations in lattice-normed spaces

We shall consider the algorithm studied in §5 for the case of operator equations in a space R metrized by elements of G. Recall that this algorithm pertains to the equation

$$x = F(x, x) \tag{8.1}$$

in some space E; the approximations x_n to the solutions of this equation are defined as solutions of the equations

$$x_n = F_k(x_n, x_{n-1}) \quad (x_0 \in E, \quad n = 1, 2, \dots), \tag{8.2}$$

where the operators F_k are defined by

$$\begin{gathered} F_1(x, y) = F(x, y), \\ F_i(x, y) = F[x, F_{i-1}(x, y)] \qquad (i = 2, 3, \dots, k). \end{gathered} \tag{8.3}$$

1. Let us assume that E is lattice-normed by some archimedean K-lineal N, and that on some subset D of E the operator $F(x, y)$ satisfies the condition

$$\| F(x, y) - F(z, t) \| \leqslant M \| x - z \| + L \| y - t \|, \tag{8.4}$$

where M and L are positive linear operators defined on N.

THEOREM 8.1. *Suppose each equation* (8.2) *has a solution x_n in D, and the series $\Sigma_{i=0}^{\infty} \Delta_k^i a$ and $\Sigma_{i=0}^{\infty} E_k^i a$, where*

$$\Delta_k = \sum_{i=0}^{k-1} L^i M, \quad E_k = (I - \Delta_k)^{-1} L^k,$$

are uniformly convergent for any $a \in N$.

Then the sequence $\{x_n\}$ converges to the unique solution in D of the equation

$$x = F_k(x, x) \tag{8.5}$$

for any $x_0 \in D$, *and the following error estimated holds*:

$$\|x - x_n\| \leqslant (I - E_k)^{-1} E_k^{n-p+1} \|x_p - x_{p-1}\| \qquad (1 \leqslant p \leqslant n). \tag{8.6}$$

PROOF. In view of (8.3), it follows from (8.2) that

$$\begin{aligned}\|x_n - x_{n-1}\| &= \|F_k(x_n, x_{n-1}) - F_k(x_{n-1}, x_{n-2})\| \\ &= \|F(x_n, F_{k-1}(x_n, x_{n-1})) - F(x_{n-1}, F_{k-1}(x_{n-1}, x_{n-2}))\| \\ &\leqslant M\|x_n - x_{n-1}\| + L\|F_{k-1}(x_n, x_{n-1}) - F_{k-1}(x_{n-1}, x_{n-2})\| \\ &\leqslant \Delta_k\|x_n - x_{n-1}\| + L^k\|x_{n-1} - x_{n-2}\|.\end{aligned}$$

Since $\Sigma_{i=0}^{\infty} \Delta_k^i a$ is uniformly convergent for any $a \in N$, the inverse $(I - \Delta_k)^{-1}$ exists, and we deduce from the above inequality that

$$\|x_n - x_{n-1}\| \leqslant E_k \|x_{n-1} - x_{n-2}\| \leqslant E_k^{n-p} \|x_p - x_{p-1}\| \qquad (1 \leqslant p \leqslant n-1).$$

Next, proceeding as in §6.3 with E_k in place of P (see (6.19)), one proves that the sequence $\{x_n\}$ converges to a limit $x \in D$ and satisfies (8.6).

We now show that x is a solution of (8.5). We have

$$\begin{aligned}\theta \leqslant \|F_k(x, x) - x\| &\leqslant \|F_k(x, x) - F_k(x_n, x_{n-1})\| + \|x_n - x\| \\ &\leqslant \Delta_k\|x - x_n\| + L^k\|x - x_{n-1}\| + \|x_n - x\|.\end{aligned}$$

Hence, since the operators Δ_k and L^k are continuous and $\lim_{n\to\infty} \|x_n - x\| = \theta$, it follows that $x = F_k(x, x)$.

The uniqueness of the solution of (8.5) is proved in the usual way. Assuming the existence of a solution $x' \in D$ of (8.5) other than x, we obtain

$$\|x - x'\| = \|F_k(x, x) - F_k(x', x')\| \leqslant \Delta_k\|x - x'\| + L^k\|x - x'\|,$$

whence $\|x - x'\| \leqslant E\|x - x'\|$. Repeating the argument n times, we obtain

$$\|x - x'\| \leqslant E^n \|x - x'\|.$$

But since $\Sigma_0^{\infty} E^k a$ is uniformly convergent for any $a \in N$ we have

$$\lim_{\infty} E^n \|x - x'\| = \theta,$$

and so this inequality is possible only if $x = x'$.

Had we chosen any other sequence $\{x_n'\}$ of solutions of equations (8.2) in D (provided another sequence exists), similar arguments would have shown that this sequence has a limit x' satisfying (8.5); but since (8.5) is uniquely solvable on D it would follow that $x' = x$, so that the sequence converges to x.

We have proved that $\{x_n\}$ converges to a solution of equation (8.5) rather than a solution of the original equation (8.1). However, since any solution of (8.1) is also a solution of (8.5), if we assume that (8.1) is solvable on D we may then conclude that the solution is unique and is therefore the limit of the sequence

$\{x_n\}$, and the error estimate (8.6) is valid.

2. We have assumed that each of equations (8.2) has a solution $x_n \in D$. Under the assumptions of Theorem 8.1, this assumption may be replaced by the condition that the operator $F_k(x, y)$ maps D into itself, $F_k(x, y) \in D$ for $x, y \in D$. This will be the case, for example, if $F(x, y)$ maps D into itself.

It is easy to see that if $F_k(x, y) \in D$ for all $x, y \in D$ and the other assumptions of Theorem 8.1 are valid, then equation (8.2) may be solved for each n by the usual method of successive approximations

$$x_n^{(m)} = F_k\left(x_n^{(m-1)}, x_{n-1}\right), \tag{8.7}$$

and this also yields an existence and uniqueness proof for the solution x_n of equation (8.2) on D.

We now consider the question of determining the radius of a ball $S(z, r)$ (if such exists) which the operator $F(x, y)$ (hence also $F_k(x, y)$) maps into itself.

Suppose that for all $x, y \in S(z, r)$, i.e., $\|x - z\|, \|y - z\| \leqslant r$, where r is to be determined, we have

$$\| F(x, y) - z \| \leqslant V(r), \tag{8.8}$$

where $V(r)$ is a known function with domain and range in N. If there exists $r^* \in N$ such that

$$V(r^*) \leqslant r^*, \tag{8.9}$$

then clearly $F(x, y) \in S(z, r^*)$ for $x, y \in S(z, r^*)$.

The radius of the required ball $S(z, r)$ may be defined as the element $r^* \in N$ satisfying the condition

$$\| F(z, z) - z \| + W(r) \leqslant r, \tag{8.10}$$

where $W(r)$ is determined by

$$\| F(x, y) - F(z, z) \| \leqslant W(r) \tag{8.11}$$

for all $x, y \in S(z, r)$. Then, as in the previous case, $F(x, y) \in S(z, r^*)$ for $x, y \in S(z, r^*)$.

As the operator $W(r)$ one can obviously take $[M(r) + L(r)]\, r$, where $M(r)$ and $L(r)$, which generally depend on r, are determined by the inequality

$$\| F(x, y) - F(z, z) \| \leqslant M(r) \| x - z \| + L(r) \| y - z \| \tag{8.12}$$

for $x, y \in S(z, r)$. Indeed, under these conditions, if $x, y \in S(z, r^*)$, then

$$\begin{aligned} \| F(x, y) - z \| &\leqslant \| F(x, y) - F(z, z) \| + \| F(z, z) - z \| \\ &\leqslant M(r^*) \| x - z \| + L(r^*) \| y - z \| + \| F(z, z) - z \| \\ &\leqslant M(r^*) r^* + L(r^*) r^* + \| F(z, z) - z \| \leqslant r^*, \end{aligned}$$

i.e., $F(x, y) \in S(z, r^*)$.

3. We consider some special cases of the operator $F(x, y)$. Let

$$F(x, y) = Rx + Sy, \tag{8.13}$$

where R and S are operators from E to E, and S is linear. Then it is readily seen that

$$F_k(x, y) = \sum_{i=0}^{k-1} S^i Rx + S^k y. \tag{8.14}$$

If the operator $\Sigma_0^{k-1} S^i R$ has a resolvent operator R_k, the successive approximation process defined by (8.2) reduces to the classical process for the equation

$$x = R_k S^k x, \tag{8.15}$$

which is equivalent to (8.5) and in this case has the form

$$x = \sum_{i=0}^{k-1} S^i Rx + S^k x. \tag{8.16}$$

If the operator R in (8.13) is singular,

$$Rx = \sum_{i=1}^{l} \Phi_i(x) \varphi_i + f, \tag{8.17}$$

where φ_i and f are elements of E and the Φ_i are functionals on E and l is a natural number, then the operator $\Sigma_0^{k-1} S^i R$ is also singular. Indeed,

$$\sum_{i=0}^{k-1} S^i Rx = \sum_{j=0}^{k-1} S^j \left(\sum_{i=1}^{l} \Phi_i(x) \varphi_i + f \right) = \sum_{i=1}^{l} \Phi_i(x) \psi_{i,k} + f_k,$$

where

$$\psi_{i,k} = \sum_{j=0}^{k-1} S^j \varphi_i, \quad f_k = \sum_{j=0}^{k-1} S^j f.$$

Consequently, to determine the successive approximations x_n we have equations with a singular operator

$$x_n = \sum_{i=0}^{l} \Phi_i(x_n) \psi_{i,k} + f_k + S^k x_{n-1}, \tag{8.18}$$

whose solution reduces at each step to solution of systems of algebraic or transcendental equations of order at most l.

REMARK. Since any space E lattice-normed by an archimedean K-lineal is in particular a space metrized by elements of a set G, the convergence criteria and error estimates of the generalized iterative method established in §5 are also valid in the present case. The funcuons $\varphi(c, d)$, $\varphi_k(c, d)$, $B_{n,k}(b)$ and $\psi_n(u)$ are in this case

$$\varphi(c, d) = Mc + Ld, \quad \varphi_k(c, d) = \Delta_k c + L^k d, \qquad B_{n,k}(b) = \mathrm{E}_k^n b,$$

$$\psi_n(u) = \| x_{n+1} - x_n \| + \Delta_k \mathrm{E}^{n+1} b + L^k u.$$

§9. Generalized algorithms for the method of averaged functional corrections

In this section we study the generalized iterative method (8.2) for equations in a lattice-normed space when the operator $F(x, y)$ is specially chosen.

Consider the operator equation

$$x = Tx, \tag{9.1}$$

where T is defined on a space E lattice-normed by an archimedean K-lineal N. We assume that the equation has a solution in E.

Let P be a linear projection ($P = P^2$) mapping E onto some subspace E_P. Let Q denote the difference $I - P$, where I is the identity operator.

1. We consider three iterative projection methods, corresponding to different forms of the operator $F(x, y)$ in terms of the original operator T:

a) $F(x, y) = PTx + QTy$,

b) $F(x, y) = T(Px + Qy)$,

c) $F(x, y) = PTx + QT(Px + Qy)$.

It is clear that for each of these cases $F(x, x) = Tx$.

In case (a), as is readily seen, the operator $F_k(x, y)$ is

$$F_k(x, y) = R_k(PTx, QTy), \tag{9.2}$$

where the operators $R_k(PTx, QTy)$ are defined recursively by

$$\begin{gathered} R_1(PTx, QTy) = PTx + QTy, \\ R_i(PTx, QTy) = PTx + QTR_{i-1}(PTx, QTy) \quad (i = 2, 3, \ldots, k), \end{gathered} \tag{9.3}$$

and the successive approximations to the solution of equation (9.1) are the solutions of the equations

$$u_n = R_k(PTu_n, QTu_{n-1}) \, (u_0 \in E, \; n = 1, 2, \ldots). \tag{9.4}$$

In case (b), $F_k(x, y)$ is defined by

$$F_k(x, y) = S_k(Px, Qy), \tag{9.5}$$

where the operators $S_k(Px, Qy)$ are defined by

$$\begin{gathered} S_1(Px, Qy) = T(Px + Qy), \\ S_i(Px, Qy) = T[Px + QS_{i-1}(Px, Qy)] \qquad (i = 2, 3, \ldots, k). \end{gathered} \tag{9.6}$$

To evaluate the successive approximations v_n in this case, we have the operator equations

$$v_n = S_k(Pv_n, Qv_{n-1}) \quad (v_0 \in E, \quad n = 1, 2, \ldots). \tag{9.7}$$

Finally, the operators $F_k(x, y)$ for case (c) have the form

$$F_k(x, y) = U_k(PTx, Px, Qy), \tag{9.8}$$

where

$$\begin{aligned} U_1(PTx, Px, Qy) &= PTx + QT(Px + Qy), \\ U_i(PTx, Px, Qy) &= PTx + QT[Px + QU_{i-1}(PTx, Px, Qy)] \\ &\quad (i = 2, 3, \ldots, k), \end{aligned} \tag{9.9}$$

and the successive approximations w_n are defined by the equations

$$w_n = U_k(PTw_n, Pw_n, Qw_{n-1}) \, (w_0 \in E, \quad n = 1, 2, \ldots). \tag{9.10}$$

As mentioned previously, the operator $F(x, y)$ should be constructed (if possible) so that the equation

$$x = F_k(x, y) \tag{9.11}$$

is easily solved for x and the principal part of $F_k(x, y)$ is a function of x (i.e., the operator $F_k(x, y)$ should be "weakly dependent" on y). It is evident from formulas (a), (b) and (c) for the operator $F(x, y)$ that if P is close to the identity operator the desired situation will have been achieved in many cases. We shall now show that in each of the three cases solution of the corresponding equation (9.11) may be reduced to solution of a certain equation in the subspace E_P and to evaluation of certain operators. In the first case, applying the operator PT to both sides of (9.4), we get an equation

$$PTu_n = PTR_k(PTu_n, QTu_{n-1})$$

for PTu_n, which is an element of E_P. Determining PTu_n from this equation and substituting it into the right-hand side of (9.4), we obtain the required element u_n. In the second case it is sufficient to apply P to both sides of (9.7), to obtain the equation

$$Pv_n = PS_k(Pv_n, Qv_{n-1})$$

for Pv_n, which is again an element of E_P. Substituting the resulting value of Pv_n into the right-hand side of (9.7), we find v_n. In the third case we use the fact that $PQ = \theta$ and $Pw_n = PTw_n$, and, applying the operator PT for equation (9.10), obtain an equation

$$PTw_n = PTU_k(PTw_n, PTw_n, Qw_{n-1})$$

for the elements PTw_n of E_P. Substituting the solution PTw_n into (9.10), we find the required element w_n.

We have only demonstrated that equation (9.11) may be reduced in these cases to solution of a certain new equation in E. The actual solution of equations (9.4), (9.7) and (9.10) need not follow the methods described here–in many cases some other method may prove more effective, depending on the specific problem at hand.

If the operator T in (9.1) has the form $Tx = f + Ax$, where A is a linear operator in E, the respective expressions for $F(x, y)$ in the above cases are

a) $F(x, y) = f + PAx + QAy$,

b) $F(x, y) = f + APx + AQy$,

c) $F(x, y) = f + PAx + QAPx + QAQy$,

and the equations for u_n, v_n and w_n (equations (9.4), (9.7) and (9.10), respectively) reduce to (7.18), (7.19) and (7.20).

The above methods are related to the well-known method of Schmidt, first applied to linear integral equations and subsequently to operator equations [161]. For the linear operator equation

$$x = f + Ax \tag{9.12}$$

where A is a compact operator, the idea of this method is to approximate A by some finite rank operator B such that $\|C\| < 1$, where $C = A - B$, and to replace (9.12) by the equivalent equation

$$x = (I - C)^{-1}f + (I - C)^{-1}Bx, \tag{9.13}$$

where the operator $(I - C)^{-1}B$ is finite rank (note that $(I - C)^{-1}$ exists since $\|C\| < 1$). And since solution of equations with finite rank operators reduces to solution of a finite system of linear algebraic equations, investigation of (9.12) thus becomes a relatively simple matter.

In each of cases (a), (b) and (c), the operator A is also approximated by a finite rank operator: PA, AP and $PA + QAP$. Since it is generally impossible to determine $(I - C)^{-1}$, we confine ourselves to the approximating operator Σ_0^{k-1} $\Sigma_0^{k-1}C^i$. Setting $x_0 = \theta$, we see that the first approximation to the equation is

$$x_1 = \sum_{i=0}^{k-1} C^i f + \sum_{i=0}^{k-1} C^i Bx_1,$$

which tends to equation (9.13) as $k \to \infty$. Thus, as $k \to \infty$ this equation approaches the equation of Schmidt's method, and its solution–the first approximation to the exact solution of the original equation–approaches the solution of the original equation. Restricting attention to a finite value of k, we construct an iterative process with respect to n. This makes it possible to use the approximate solution in constructing the next, improved approximation, whereas when k is increased the previous approximation is not used in the construction of the next.

2. We now establish some sufficient convergence conditions and error estimates for the successive approximations defined by our iterative projection methods.

Let us first assume that equations (9.4), (9.7) and (9.10) are solvable for u_n, v_n and w_n, so that

$$u_n = RQTu_{n-1}, \tag{9.14}$$

$$v_n = SQv_{n-1}, \tag{9.15}$$

$$w_n = UQw_{n-1}, \tag{9.16}$$

where R, S and U are certain operators. We thus have the classical method of successive approximations applied to new equations, equivalent to the equation

$$x = F_k(x, x). \tag{9.17}$$

Of course, in the general case no expressions for the operators R, S and U are available, but in many cases one can find estimates for their Lipschitz operators, which are needed in convergence investigations. Using the convergence criteria of usual method of successive approximations, we obtain convergence criteria for our new methods. For example, the processes (9.14), (9.15) and (9.16), and hence also the equivalent processes (9.4), (9.7) and (9.10), respectively, will converge provided the series

$$\sum_{i=0}^{\infty} L_{QTRQ}^{i} a, \quad \sum_{i=0}^{\infty} L_{QSQ}^{i} a \text{ and } \sum_{i=0}^{\infty} L_{QUQ}^{i} a,$$

where L denotes a Lipschitz operator for the appropriate operator, are uniformly convergent for any $a \in N$. We then have the following error estimates:

$$\| x - u_n \| \leqslant L_{RQ} (I - L_{QTRQ})^{-1} L_{QTRQ}^{n-p} \| QTu_p - QTu_{p-1} \| \quad (1 \leqslant p \leqslant n), \tag{9.18}$$

$$\| x - v_n \| \leqslant L_{SQ} (I - L_{QSQ})^{-1} L_{QSQ}^{n-p} \| Q (v_p - v_{p-1}) \| \quad (1 \leqslant p \leqslant n), \tag{9.19}$$

$$\| x - w_n \| \leqslant L_{UQ} (I - L_{QUQ})^{-1} L_{QUQ}^{n-p} \| Q (w_p - w_{p-1}) \| \quad (1 \leqslant p \leqslant n). \tag{9.20}$$

To prove this assertion, it suffices to observe that under these conditions the sequences $\{ QTu_n \}$, $\{ Qv_n \}$ and $\{ Qw_n \}$ are Cauchy, and since

$$\|u_n - u_{n-1}\| \leqslant L_{RQ}\|QTu_{n-1} - QTu_{n-2}\|,$$
$$\|v_n - v_{n-1}\| \leqslant L_{SQ}\|Qv_{n-1} - Qv_{n-2}\|,$$
$$\|w_n - w_{n-1}\| \leqslant L_{UQ}\|Qw_{n-1} - Qw_{n-2}\|,$$

so are the sequences $\{u_n\}, \{v_n\}$ and $\{w_n\}$.

If equation (9.1) is linear, $Tx = f + Ax$, where f is an element of E and A a linear operator, we may write (9.14), (9.15) and (9.16) thus:

$$u_n = \left[I - PA - \sum_{i=0}^{k-2} (QAQ)^i QAPA\right]^{-1} \left\{\left[I + \sum_{i=0}^{k-2} (QAQ)^i QA\right] f + (QAQ)^{k-1} QAu_{n-1}\right\}, \tag{9.21}$$

$$v_n = \left[I - AP - AQ\sum_{i=0}^{k-2} (QAQ)^i AP\right]^{-1} \left\{\left[I + AQ\sum_{i=0}^{k-2} (QAQ)^i\right] f + AQ\,(QAQ)^{k-1} Qv_{n-1}\right\}, \tag{9.22}$$

$$w_n = \left[I - \sum_{i=0}^{k-1} (QAQ)^i\,(PA + QAP)\right]^{-1} \left[\sum_{i=0}^{k-1} (QAQ)^i f + (QAQ)^k Qw_{n-1}\right]. \tag{9.23}$$

Note that if v_0 is put equal to Tu_0, where u_0 is an arbitrary element of E, then the successive approximation u_n and v_n are related by $v_n = Tu_n$, $n = 1, 2, \ldots$. Indeed, denoting

$$z_n = Pv_n + QS_{k-1}(Pv_n, Qv_{n-1}),$$

we have $v_n = Tz_n$. Substituting this value for v_n into the preceding formula, we get

$$z_n = PTz_n + QTR_{k-1}(PTz_n, QTz_{n-1}),$$

i.e., $z_n = R_k(PTz_n, QTz_{n-1})$, and this is precisely the equation (9.4) for u_n. Consequently, if $z_0 = u_0$, then $z_n = u_n$ $(n = 1, 2, \ldots)$. Thus, if the successive approximations u_n converge to a solution of the equation, the same is true for the v_n.

To establish convergence conditions for algorithms (9.4) and (9.7), we can use the results of §8 for the general algorithm. On this basis one proves the following assertion.

THEOREM 9.1. *Let PT and QT be operators in E such that*

$$\|PTx - PTy\| \leqslant L_{PT}\|x-y\|, \tag{9.24}$$
$$\|QTx - QTy\| \leqslant L_{QT}\|x-y\|, \tag{9.25}$$

and suppose that if L_{PT} and L_{QT} are Lipschitz operators and

$$\Delta_k = \sum_{i=0}^{k-1} L_{QT}^i L_{PT}, \quad \mathrm{E}_k = (I-\Delta_k)^{-1} L_{QT}^k,$$

then the series $\Sigma_{i=0}^{\infty}\Delta_k^i a$ and $\Sigma_{i=0}^{\infty}\mathrm{E}_k^i a$ are uniformly convergent for any $a \in N$.

Then equations (9.4) *and* (9.7) *are uniquely solvable for u_n and v_n for any k, n and $u_0 \in E$, $v_0 \in E$, and the sequences $\{u_n\}$ and $\{v_n\}$ converge to the (unique) solution of equation* (9.1). *The following estimates hold:*

$$\|x-u_n\| \leqslant (I-\mathrm{E}_k)^{-1}\mathrm{E}_k^{n-p+1}\|u_p - u_{p-1}\| \qquad (1 \leqslant p \leqslant n), \tag{9.26}$$
$$\|x-v_n\| \leqslant L_T(I-\mathrm{E}_k)^{-1}\mathrm{E}_k^{n-p+1}\|z_p - z_{p-1}\| \qquad (2 \leqslant p \leqslant n), \tag{9.27}$$
$$\|x-v_n\| \leqslant L_T(I-\mathrm{E}_k)^{-1}\mathrm{E}_k^{n-p}(I-\Delta_k)^{-1}L_{QT}^{k-1}\|Q(v_p - v_{p-1})\| \quad (1 \leqslant p \leqslant n), \tag{9.28}$$

where L_T is a Lipschitz operator for T.

PROOF. That equations (9.4) are uniquely solvable for u_n and the solutions converge to a solution of the equation

$$x = F_k(x, x)$$

follows directly from Theorem 8.1. Under the assumptions of our theorem, the latter equation possesses a unique solution. Hence, if (9.1) has a solution, it is unique (see §4), and therefore the sequence $\{u_n\}$ converges to this solution. The error estimate (9.26) is then precisely (8.6), which was proved in §8.

To prove that equations (9.7) are uniquely solvable for v_n and that the sequence $\{v_n\}$ converges to a solution of (9.1), we need only note that $v_n = Tz_n$. This yields also the error estimate (9.27). To verify (9.28), we simply note that

$$\|z_p - z_{p-1}\| \leqslant (I-\Delta_k)^{-1}L_{QT}^{k-1}\|Q(v_{p-1} - v_{p-2})\| \qquad (p = 2,\ 3,\ \ldots).$$

We now estalbish convergence criteria and error estimates for the algorithm (9.10).

THEOREM 9.2. *Let the operators PT and QT satisfy conditions* (9.24) *and* (9.25); *let L_{PT} and L_{QT} be suitable Lipschitz operators, set*

$$\overline{\Delta}_k = \sum_{i=1}^{k} L_{QT}^i(I-L_{PT})^{-1}L_{PT}, \quad \overline{\mathrm{E}}_k = (I-\overline{\Delta}_k)^{-1}L_{QT}^k, \quad \Delta_k = \sum_{i=0}^{k} L_{QT}^i L_{PT},$$

and assume that the series

$$\sum_{i=0}^{\infty} L_{PT}^i a, \quad \sum_{i=0}^{\infty} \overline{\Delta}_k^i a, \quad \sum_{i=0}^{\infty} \Delta_k^i a, \quad \sum_{i=0}^{\infty} \overline{E}_k^i,$$

are uniformly convergent for any $a \in N$.

Then equations (9.10) *are uniformly solvable for w_n for any $w_0 \in E$, and the sequence $\{w_n\}$ converges to a solution of equation* (9.1); *moreover the following error estimate holds*:

$$\|x - w_n\| \leqslant (I - L_{PT})^{-1} (I - \overline{E}_k)^{-1} \overline{E}_k^{n-p+1} \|Q(w_p - w_{p-1})\| \quad (1 \leqslant p \leqslant n). \tag{9.29}$$

PROOF. That equations (9.10) are uniquely solvable for w_n follows from the uniform convergence of $\Sigma_{i=0}^{\infty} \Delta_k^i a$, since by virtue of this condition the usual method of successive approximations is applicable to equations (9.10).

Using the expression for U_k and condition (9.25), and setting $\delta_n = w_n - w_{n-1}$, we deduce from (9.10) that

$$\|Q\delta_n\| \leqslant \sum_{i=1}^{k} L_{QT}^i \|P\delta_n\| + L_{QT}^k \|Q\delta_{n-1}\|. \tag{9.30}$$

Since $w_n = Pw_n + Qw_n$ and by (9.10) $Pw_n = PTw_n$, it follows from (9.24) that

$$\|P\delta_n\| \leqslant L_{PT} (\|P\delta_n\| + \|Q\delta_n\|),$$

whence

$$\|P\delta_n\| \leqslant (I - L_{PT})^{-1} L_{PT} \|Q\delta_n\|.$$

Substituting this estimate for $\|P\delta_n\|$ into the right-hand side of (9.30), we obtain

$$\|Q\delta_n\| \leqslant \overline{\Delta}_k \|Q\delta_n\| + L_{QT}^k \|Q\delta_{n-1}\|$$

and, since the uniform convergence of $\Sigma_{i=0}^{\infty} \overline{\Delta}_k^i a$ implies the existence of a positive inverse $(I - \overline{\Delta}_k)^{-1}$, we have

$$\|Q\delta_n\| \leqslant (I - \overline{\Delta}_k)^{-1} L_{QT}^k \|Q\delta_{n-1}\| = \overline{E}_k \|Q\delta_{n-1}\|.$$

Now, proceeding as in the usual method of successive approximations, one proves that $\{Qw_n\}$ is a Cauchy sequence. And since $\delta_n = P\delta_n + Q\delta_n$, and by (9.10) $P\delta_n = PTw_n - PTw_{n-1}$, it follows that

$$\|\delta_n\| \leqslant \|PTw_n - PTw_{n-1}\| + \|Q\delta_n\| \leqslant L_{PT} \|\delta_n\| + \|Q\delta_n\|.$$

Hence $\|\delta_n\| \leqslant (I - L_{PT})^{-1} \|Q\delta_n\|$ and so $\{w_n\}$ is also a Cauchy sequence. The estimate (9.29) is now proved in the usual way (see §8).

3. In the last subsection we assumed that the conditions held throughout the space E. We now generalize, assuming that the conditions hold only in some subset D of E.

We first consider algorithm (9.4). It is clear from the definition of R_k that if $F(x, y) \in D$ for any x, y in D, i.e.,

$$PTx + QTy \in D, \tag{9.31}$$

then also $R_i(x, y) \in D$ for any $i = 1, 2, \ldots$ Consequently, if (9.31) is true for any $x, y \in D$ and the assumptions of Theorem 9.1 are satisfied in D, then equations (9.4) are uniquely solvable for u_n in D, for any $u_0 \in D$, and the sequence $\{u_n\}$ converges to the (unique) solution in D of (9.1); moreover, the error estimate (9.26) holds.

Let us illustrate a few cases in which condition (9.31) holds for $x, y \in D$.

I. Let D be a ball $S(z, r)$, where $r \in N$ and the operators PT and QT satisfy conditions (9.24) and (9.25) for all elements of the ball. Then, if

$$\|Tz - z\| + (L_{PT} + L_{QT})r \leqslant r \tag{9.32}$$

we have $PTx + QTy \in S(z, r)$ whenever $x, y \in S(z, r)$. Indeed, for any $x, y \in S(z, r)$,

$$\begin{aligned} \|PTx + QTy - z\| &= \|PTx - PTz + QTy - QTz + PTz + QTz - z\| \\ &\leqslant \|PTx - PTz\| + \|QTy - QTz\| + \|Tz - z\| \end{aligned}$$

$$\leqslant L_{PT}\|x - z\| + L_{QT}\|y - z\| + \|Tz - z\| \leqslant \|Tz - z\| + (L_{PT} + L_{QT})r \leqslant r,$$

so that $PTx + QTy \in S(z, r)$.

II. Suppose that in $S(z, r)$, where r is an as yet undetermined element of N, we have the estimates

$$\begin{aligned} \|PTx - Pz\| &\leqslant V_{PT}(r), \\ \|QTy - Qz\| &\leqslant V_{QT}(r), \end{aligned} \tag{9.33}$$

where $V_{PT}(r)$ and $V_{QT}(r)$ are functions with domains and ranges in N. If there is an element r^* such that

$$V_{PT}(r) + V_{QT}(r) \leqslant r, \tag{9.34}$$

then clearly $PTx + QTy \in S(z, r^*)$ if $x, y \in S(z, r^*)$.

III. Suppose that for any elements of $S(z, r)$, where r is as yet undetermined, and

$$\|PTx + QTy - PTz - QTz\| \leqslant W_{PT}(r)\|x - z\| + W_{QT}(r)\|y - z\|, \tag{9.35}$$

where the W's are operators from N to N.

If there is an element $r^* \in N$ such that

$$\|Tz - z\| + W_{PT}(r)r + W_{QT}(r)r \leqslant r, \tag{9.36}$$

then $PTx + QTy \in S(z, r^*)$ for $x, y \in S(z, r^*)$.

The proof follows the same lines as that of the analogous assertion in case I.

We now consider algorithms (9.7) and (9.10).

Analyzing the formula for R_i, we readily see that if for all x, y in some subset D of E

$$PTx + Qv_0 \in D, \tag{9.37}$$

$$PTx + QTy \in D, \tag{9.38}$$

then

$$PTx + QTR_{i-1}(PTx + Qv_{n-1}) \in D \tag{9.39}$$

for all $n = 1, 2, \ldots$ and $i = 1, 2, \ldots$ Hence, in view of the fact that $v_n = Tz_n$, we obtain the following proposition.

If D is a closed set in which the assumptions of Theorem 9.1 are satisfied, then the equations

$$\begin{aligned} z_1 &= R_k(PTz_1, Qv_0), \\ z_n &= R_k(PTz_n, QTz_{n-1}) \qquad (n = 2, 3, \ldots) \end{aligned} \tag{9.40}$$

are uniquely solvable in D for z_n $(n = 1, 2, \ldots)$ and the successive approximations $z_n \in D$ and $v_n \in D$ defined by (9.40) and (9.7) converge to the (unique) solution in D of (9.1); the error estimate (9.27) is also valid.

The proof is analogous to that of Theorem 9.1, except that instead of the whole space E one consideres only the subset D.

The situation is analogous with regard to algorithm (9.10). If (9.38) is true for any $x, y \in D$ and

$$PTx + Qw_0 \in D, \tag{9.41}$$

then

$$U_i(PTx, Px, Qw_{n-1}) \in D \tag{9.42}$$

for all $n = 1, 2, \ldots$ and $i = 1, 2, \ldots$ Thus, if D is closed and the assumptions of Theorem 9.2 hold there, then equations (9.10) are uniquely solvable in D for w_n, $n = 1, 2, \ldots$, and the solutions $\{w_n\}$ converge to the (unique) solution in D of (9.1); the error estimate (9.29) is also valid.

If the set D is a ball $S(z, r)$, conditions (9.37) and (9.38), as well as the analogous conditions (9.41) and (9.38) with v_0 replaced by w_0, will be satisfied, for example, in cases I–III above, except that instead of (9.32), (9.34) and (9.36) we have the following systems of inequalities:

1) $$\begin{aligned} &\|Qv_0 + PTz - z\| + L_{PT}\, r \leq r, \\ &\|Tz - z\| + (L_{PT} + L_{QT})\, r \leq r; \end{aligned} \tag{9.43}$$

2) $$\begin{aligned} &\|Q(v_0 - z)\| + v_{PT}(r) \leq r, \\ &V_{PT}(r) + V_{QT}(r) \leq r; \end{aligned} \tag{9.44}$$

$$3)\quad \begin{aligned} &\|Qv_0 + PTz - z\| + w_{PT}(r)\, r \leqslant r, \\ &\|Tz - z\| + [W_{PT}(r) + W_{QT}(r)]\, r \leqslant r. \end{aligned} \tag{9.45}$$

§10. A special case of operator equations and projection operators

Let us assume that the space E, the operator T of (9.1) and the projection operator P in algorithms (9.4), (9.7) and (9.10) have the property that, for all x, y, $\bar{x}$, $\bar{y} \in E$,

$$\|T(Px + Qy) - T(P\bar{x} + Q\bar{y})\| \leqslant L_{TP}\|P(x - \bar{x})\| + L_{TQ}\|Q(y - \bar{y})\|, \tag{10.1}$$

where L_{TP} and L_{TQ} are positive linear operators, and that an analogous inequality is true for the operators PT and QT, with L_{TP} and L_{TQ} replaced by L_{PTP}, L_{PTQ}, L_{QTP} and L_{QTQ}.

Condition (10.1) holds for a fairly broad class of operators. For example, it holds for bounded linear operators, and also for certain classes of nonlinear operators in direct product spaces. Special classes of such operators will be considered in Chapters II and III.

We now establish some simple convergence criteria for algorithms (9.4), (9.7) and (9.10) on the assumption that condition (10.1) is satisfied and equation (9.1) solvable.

THEOREM 10.1. *Let condition* (10.1) *and the analogous condition for the operators PT and QT be satisfied. Suppose that the series $\Sigma_{i=0}^{\infty} \tilde{\Delta}_k^i a$ and $\Sigma_{i=0}^{\infty} \tilde{Ɛ}_k^i a$, where*

$$\tilde{\Delta}_k = L_{PTP} + L_{PTQ} \sum_{i=0}^{k-2} L_{QTQ}^i L_{QTP},$$

$$\tilde{Ɛ}_k = L_{QTQ}^k + \sum_{i=0}^{k-1} L_{QTQ}^i L_{QTP} \left(I - \tilde{\Delta}_k\right)^{-1} L_{PTQ} L_{QTQ}^{k-1},$$

are uniformly convergent for any $a \in N$.

Then equations (9.4) *and* (9.7) *are uniquely solvable for u_n and v_n in E for any n, k and $u_0 \in E$, $v_0 \in E$, and the sequences of solutions converge to a solution of equation* (9.1), *with the following error estimates*:

$$\|x - u_n\| \leqslant L_1^{(k)} \left(I - \tilde{Ɛ}_k\right)^{-1} \tilde{Ɛ}_k^{n-p} \|QTu_p - QTu_{p-1}\| \quad (1 \leqslant p \leqslant n), \tag{10.2}$$

$$\|x - v_n\| \leqslant L_2^{(k)} \left(I - \tilde{Ɛ}_k\right)^{-1} \tilde{Ɛ}_k^{n-p} \|Q(v_p - v_{p-1})\| \quad (1 \leqslant p \leqslant n), \tag{10.3}$$

where

$$L_1^{(k)} = \left(I + \sum_{i=0}^{k-2} L_{QTQ}^i L_{QTP}\right)\left(I - \widetilde{\Delta}_k\right)^{-1} L_{PTQ} L_{QTQ}^{k-1} + L_{QTQ}^{k-1},$$

$$L_2^{(k)} = \left(L_{TP} + L_{TQ} \sum_{i=0}^{k-2} L_{QTQ}^i L_{QTP}\right)\left(I - \widetilde{\Delta}_k\right)^{-1} L_{PTQ} L_{QTQ}^{k-1} + L_{TQ} L_{QTQ}^{k-1}.$$

PROOF. We first prove that equations (9.4) and (9.7) are uniquely solvable for u_n and v_n. Applying the operators PT and P to both sides of (9.4) and (9.7), respectively, we get

$$PTu_n = PTR_k(PTu_n, QTu_{n-1}), \tag{10.4}$$
$$Pv_n = PS_k(Pv_n, Qv_{n-1}). \tag{10.5}$$

Viewing these as equations for PTu_n and Pv_n, we see that by virtue of the uniform convergence of $\Sigma_{i=0}^\infty \widetilde{\Delta}_k^i a$ for any $a \in N$ their right-hand sides satisfy the conditions for the usual method of successive approximations, since

$$\| PTR_k(PTx, QTu_{n-1}) - PTR_k(PTy, QTu_{n-1}) \| \leqslant \widetilde{\Delta}_k \| x - y \|,$$
$$\| PS_k(Px, Qv_{n-1}) - PS_k(Py, Qv_{n-1}) \| \leqslant \widetilde{\Delta}_k \| x - y \|$$

(the last inequalities are easily proved using the expressions for R_i and S_i and condition (10.1)). Consequently, (10.4) and (10.5) are uniquely solvable for PTu_n and Pv_n. Substituting these values of PTu_n and Pv_n into the right-hand sides of (9.4) and (9.7), we obtain u_n and v_n. Since PTu_n and Pv_n are uniquely determined, the same is true of u_n and v_n.

We now prove that the successive approximations u_n and v_n converge to a solution of (9.1). Set $z_n - z_{n-1} = \delta z_n$. From (9.4) and (9.7), we find that

$$Q\delta Tu_n = QTR_k(PTu_n, QTu_{n-1}) - QTR_k(PTu_{n-1}, QTu_{n-2}) \quad (n = 2, 3, \ldots),$$

$$Q\delta v_n = QS_k(Pv_n, Qv_{n-1}) - QS_k(Pv_{n-1}, Qv_{n-2}) \qquad (n = 2, 3, \ldots),$$

whence, by inequality (10.1),

$$\| Q\delta Tu_n \| \leqslant \sum_{i=0}^{k-1} L_{QTQ}^i L_{QTP} \| P\delta Tu_n \| + L_{QTQ}^k \| Q\delta Tu_{n-1} \|, \tag{10.6}$$

$$\| Q\delta v_n \| \leqslant \sum_{i=0}^{k-1} L_{QTQ}^i L_{QTP} \| P\delta v_n \| + L_{QTQ}^k \| Q\delta v_{n-1} \|. \tag{10.7}$$

Similarly, we obtain

$$\| P\delta Tu_n \| \leqslant \widetilde{\Delta}_k \| P\delta Tu_n \| + L_{PTQ} L_{QTQ}^{k-1} \| Q\delta Tu_{n-1} \|,$$
$$\| P\delta v_n \| \leqslant \widetilde{\Delta}_k \| P\delta v_n \| + L_{PTQ} L_{QTQ}^{k-1} \| Q\delta v_{n-1} \|,$$

whence, since the uniform convergence of $\Sigma_{i=0}^\infty \widetilde{\Delta}_k^i a$ implies the existence of a positive inverse $(I - \widetilde{\Delta}_k)^{-1}$, we obtain

$$\| P\delta Tu_n \| \leqslant \left(I - \widetilde{\Delta}_k\right)^{-1} L_{PTQ} L_{QTQ}^{k-1} \| Q\delta Tu_{n-1} \|, \tag{10.8}$$

$$\| P\, \delta v_n \| \leqslant \left(I - \widetilde{\Delta}_k\right)^{-1} L_{PTQ} L_{QTQ}^{k-1} \| Q\delta v_{n-1} \|. \tag{10.9}$$

Substituting these estimates into (10.6) and (10.7), respectively, we get

$$\| Q\delta Tu_n \| \leqslant \widetilde{E}_k \| Q\delta Tu_{n-1} \|, \tag{10.10}$$

$$\| Q\delta v_n \| \leqslant \widetilde{E}_k \| Q\delta v_{n-1} \|. \tag{10.11}$$

Since $\Sigma_{i=0}^{\infty} \widetilde{E}_k^i a$ is uniformly convergent for any $a \in N$, the sequences $\{ QTu_n \}$ and $\{ Qv_n \}$ are Cauchy. We claim that $\{ u_n \}$ and $\{ v_n \}$ are also Cauchy sequences. Indeed, using the basic inequality (10.1), we easily show that

$$\| \delta u_n \| \leqslant \left(I + \sum_{i=0}^{k-2} L_{QTQ}^i L_{QTP} \right) \| P\delta Tu_n \| + L_{QTQ}^{k-1} \| Q\delta Tu_{n-1} \|,$$

$$\| \delta v_n \| \leqslant \left(L_{TP} + L_{TQ} \sum_{i=0}^{k-2} L_{QTQ}^i L_{QTP} \right) \| P\delta v_n \| + L_{TQ} L_{QTQ}^{k-1} \| Q\delta v_{n-1} \|.$$

Substituting (10.8) and (10.9) into the right-hand sides of these inequalities we obtain

$$\| \delta u_n \| \leqslant L_1^{(k)} \| Q\delta Tu_{n-1} \|, \tag{10.12}$$

$$\| \delta v_n \| \leqslant L_2^{(k)} \| Q\delta v_{n-1} \|. \tag{10.13}$$

Since we have proved that $\{ Q\delta\, Tu_n \}$ and $\{ Q\delta\, v_n \}$ are Cauchy sequences, it follows from (10.12) and (10.13) that $\{ u_n \}$ and $\{ v_n \}$ are also Cauchy sequences. Consequently, these sequences converge to limits u and v, respectively, which must be in E since E is complete.

We claim that u and v are solutions of the respective equations

$$x = R_k (PTx,\, QTx) \tag{10.14}$$

and

$$x = S_k (Px,\, Qx). \tag{10.15}$$

Indeed, we have the equalities

$$\begin{aligned} \| u - R_k (PTu,\, QTu) \| &\leqslant \| u - u_n \| \\ &+ \| R_k (PTu_n,\, QTu_{n-1}) - R_k (PTu,\, QTu) \| \leqslant \| u - u_n \| \\ &+ \left(I + \sum_{i=0}^{k-2} L_{QTQ}^i L_{QTP} \right) \| PTu_n - PTu \| + L_{QTQ}^{k-1} \| QTu_{n-1} - QTu \|, \end{aligned}$$

$$\| v - S_k (Pv,\, Qv) \| \leqslant \| v - v_n \| + \| S_k (Pv_n,\, Qv_{n-1}) - S_k (Pv,\, Qv) \| \leqslant$$

$$\leqslant \|v - v_n\| + \left(L_{TP} + L_{TQ} \sum_{i=0}^{k-2} L_{QTQ}^{i} L_{QTP}\right) \|P(v_n - v)\| + L_{TQ} L_{QTQ}^{k-1} \|Q(v_{n-1} - v)\|,$$

whence it follows that u and v satisfy (10.14) and (10.15), respectively.

To prove uniqueness, we assume as usual that each of equations (10.14) and (10.15) has two solutions (u, u' and v, v', respectively); proceeding as in the derivation of (10.10), (10.11) and (10.12), (10.13), we get

$$\|Q(Tu - Tu')\| \leqslant \widetilde{E}_k \|Q(Tu - Tu')\|,$$
$$\|Q(v - v')\| \leqslant \widetilde{E}_k \|Q(v - v')\|$$

and

$$\|u - u'\| \leqslant L_1^{(k)} \|Q(Tu - Tu')\|,$$
$$\|v - v'\| \leqslant L_2^{(k)} \|Q(v - v')\|,$$

whence it follows in particular that $u = u'$ and $v = v'$.

Since equation (9.1) has a solution x by assumption and each solution is also a solution of (10.14) and (10.15), it follows that $x = u$ and $x = v$.

The error estimates (10.2) and (10.3) are proved in the usual way.

The next theorem gives convergence criteria for algorithm (9.10).

THEOREM 10.2. *Suppose that condition* (10.1) *and the analogous condition for the operators PT and QT are satisfied. Suppose that the series $\Sigma_{i=0}^{\infty} \hat{\Delta}_k^i a$ and $\Sigma_{i=0}^{\infty} \hat{E}_k^i a$, where*

$$\hat{\Delta}_k = L_{PTP} + L_{PTQ} \sum_{i=0}^{k-1} L_{QTQ}^{i} L_{QTP},$$

$$\hat{E}_k = L_{QTQ}^{k} + \sum_{i=0}^{k-1} L_{QTQ}^{i} L_{QTP} \left(I - \hat{\Delta}_k\right)^{-1} L_{PTQ} L_{QTQ}^{k},$$

are uniformly convergent for any $a \in N$.

Then equations (9.10) *are uniquely solvable for w_n in E, and the sequences of solutions converges to a solution of* (9.1) *satisfying the error estimate*

$$\|x - w_n\| \leqslant L^{(k)} \left(I - \hat{E}_k\right)^{-1} \hat{E}_k^{n-p} \|Q(w_p - w_{p-1})\| \quad (1 \leqslant p \leqslant n), \tag{10.16}$$

where

$$L^{(k)} = \left(I + \sum_{i=0}^{k-1} L_{QTQ}^{i} L_{QTP}\right) \left(I - \hat{\Delta}_k\right)^{-1} L_{PTQ} L_{QTQ}^{k} + L_{QTQ}^{k}.$$

PROOF. We first show that for each n equation (9.10) has a unique solution w_n. Indeed, applying the operator PT to both sides of the equation and recalling that $PTw_n = Pw_n$, we obtain an equation

$$PTw_n = PTU_k(PTw_n,\ PTw_n,\ Qw_{n-1}) \tag{10.17}$$

for PTw_n. It is readily seen that the operator on the right of this equation satisfies the condition

$$\| PTU_k(Px,\ Px,\ Qw_{n-1}) - PTU_k(Py,\ Py,\ Qw_{n-1}) \| \leqslant \hat{\Delta}_k \| x - y \|,$$

and so we may treat equation (10.17) by the usual method of successive approximations. Since the elements PTw_n are uniquely determined from (10.17), we may substitute them in the right-hand side of (9.10), to get the unique solutions w_n.

We now prove that the resulting sequence $\{w_n\}$ converges to a solution of (9.1). To simplify the notation, we put $\delta_n = w_n - w_{n-1}$. From (10.17), using the equality $PTw_n = Pw_n$ and (10.1), we obtain

$$\| P\delta_n \| \leqslant \hat{\Delta}_k \| P\delta_n \| + L_{PTQ} L^k_{QTQ} \| Q\delta_{n-1} \|,$$

whence

$$\| P\delta_n \| \leqslant \left(I - \hat{\Delta}_k \right)^{-1} L_{PTQ} L^k_{QTQ} \| Q\delta_{n-1} \|. \tag{10.18}$$

It follows from (9.10) that

$$\| Q\delta_n \| \leqslant \sum_{i=0}^{k-1} L^i_{QTQ} L_{QTP} \| P\delta_n \| + L^k_{QTQ} \| Q\delta_{n-1} \|, \tag{10.19}$$

and then, substituting the estimate (10.18) for $\|P\delta_n\|$ on the right of (10.19), we get

$$\| Q\delta_n \| \leqslant \hat{E}_k \| Q\delta_{n-1} \|. \tag{10.20}$$

Since by assumption $\Sigma_0^\infty E_k^i\, a$ is uniformly convergent for any $a \in N$, the sequence $\{Qw_n\}$ is a Cauchy sequence. Now, since by (9.10)

$$\| \delta_n \| \leqslant \left(I + \sum_{i=0}^{k-1} L^i_{QTQ} L_{QTP} \right) \| P\delta_n \| + L^k_{QTQ} \| Q\delta_{n-1} \|,$$

it follows by substitution of the estimate (10.18) for $\|P\delta_n\|$ in the right of this inequality that

$$\| \delta_n \| \leqslant L^{(k)} \| Q\delta_{n-1} \|, \tag{10.21}$$

whence it follows, in particular, that $\{w_n\}$ is also a Cauchy sequence. Since E is a complete space, it has a limit w, which we claim is a solution of the equation

$$x = U_k(PTx,\ Px,\ Qx)\,. \tag{10.22}$$

Indeed,

$$\begin{aligned}\|w - U_k(PTw, PTw, Qw)\| \\ \leqslant \|w - w_n\| + \|U_k(PTw_n, PTw_n, Qw_{n-1}) - U_k(PTw, PTw, Qw)\| \\ \leqslant \|w - w_n\| + \left(I + \sum_{i=0}^{k-1} L_{QTQ}^{i} L_{QTP}\right) \|P(w_n - w)\| + L_{QTQ}^{k} \|Q(w_{n-1} - w)\|.\end{aligned}$$

This inequality implies that w is indeed a solution of (10.22). We now show that it is the unique solution in E. Suppose that $w' \in E$ is some other solution of (10.22). Then, proceeding as in the derivation of (10.20) and (10.21), we see that

$$\|Q(w - w')\| \leqslant \hat{E}_k \|Q(w - w')\|$$

and

$$\|w - w'\| \leqslant L^{(k)} \|Q(w - w')\|,$$

and so, by the assumptions of the theorem, $w = w'$. Finally, since (9.1) has a solution x by assumption, we have $x = w$.

The error estimate (10.16) is established in the usual way.

§11. Modified generalized iterative method for operator equations in lattice-normed spaces

The modified generalized iterative method considered in this section is a further development of a method proposed in [103] for operator equations in a Banach space.

Suppose that the operator T in the equation

$$x = Tx \tag{11.1}$$

may be expressed as $T = U + S$, where each of the operators U and S has domain and range in a space E lattice-normed by an archimedean lineal N. Setting $x = u + v$, we may write (11.1) as two operator equations for u and v:

$$\begin{aligned} u &= U(u + v), \\ v &= S(u + v). \end{aligned} \tag{11.2}$$

Indeed, adding equations (11.2) term by term, we obtain $u + v = U(u + v) + S(u + v)$, i.e., $x = Tx$. Consequently, if u and v are solutions of (11.2), then $x = u + v$ is a solution of the original equation (11.1).

We define successive approximations u_n and v_n to the solutions of (11.2) as solutions of the systems

$$\begin{aligned} u_n &= U(u_n + v_n), \\ v_n &= S_k(u_n, v_{n-1}), \end{aligned} \tag{11.3}$$

where the operators S_k are defined recursively by

$$\begin{aligned} S_1(u, v) &= S(u+v), \\ S_i(u, v) &= S[u + S_{i-1}(u, v)] \qquad (i = 2, 3, \dots, k), \end{aligned} \tag{11.4}$$

and v_0 is an arbitrary element of E.

It is then natural to take $x_n = u_n + v_n$, where u_n, v_n is a solution of system (11.3), as the nth approximation to the solution of (11.1).

In the special case $k = 1$, algorithm (11.3) reduces to the procedure considered in [103].

In many cases it is much easier to solve system (11.3) than the original equation (11.1). For example, if U is the finite rank operator

$$Ux = \sum_{i=1}^{l} \Phi_i(x)\,\varphi_i,$$

where the Φ_i are functionals, the φ_i elements of E and l is a finite natural number, then solution of system (11.3) reduces to solving a finite system of algebraic or transcendental equations of order at most l and evaluation of a few operators. Indeed, substituting the expression for v_n from the second equation of (11.3) into the first equation, we get an operator equation for u_n involving a finite rank operator. We then substitute the solution u_n into the second equation of (11.3) to find v_n and hence also x_n.

Determining the sequences $\{u_n\}$ and $\{v_n\}$ from (11.3), we thus find successive approximations to the solution of the system

$$\begin{aligned} u &= U(u+v), \\ v &= S_k(u, v), \end{aligned} \tag{11.5}$$

which may, in general, have more solutions than (11.2); thus the two systems need not be equivalent. It is clear that every solution of (11.2) is also a solution of (11.5). But the converse is generally false, as is readily seen from the following example. Let $Ux = 1 + x^2$ and $Sx = -x$ be functions of a complex variable. Then (11.2) is a system of algebraic equations:

$$\begin{aligned} u &= 1 + (u+v)^2, \\ v &= -(u+v). \end{aligned}$$

This system has a unique solution $u = 2, v = -1$. However, (11.5) is now

$$\begin{aligned} u &= 1 + (u+v)^2, \\ v &= v \end{aligned}$$

for any even k, and coincides with the original system for odd k. For even k the system has infinitely many solutions

$$v = c,\ u = \frac{1}{2}\left(1 - 2c \pm \sqrt{-3-4c}\right),$$

where c is any number.

Since any solution of system (11.2) is a solution of (11.5), these systems will be equivalent if it is known, say, that (11.2) has at least l solutions and (11.5) at most l solutions.

The convergence of the successive approximations (11.3) may be investigated by reducing them to classical successive approximations for some new equations, provided the appropriate resolvent operators exist. In fact, substituting the expression for v_n from the second equation of (11.3) into the first, we obtain

$$u_n = U\,[u_n + S_k\,(u_n,\, v_{n-1})],$$

whence, on the assumption that the latter is uniquely solvable for u_n, we obtain $u_n = \bar{R}_k v_{n-1}$, where $\bar{R}_k$ is a resolvent operator. Substituting this into the right-hand side of the second equation, we get

$$v_n = S_k\,(\bar{R}v_{n-1},\, v_{n-1}) = \bar{S}_k v_{n-1},$$

where $\bar{S}_k$ is some operator. We then determine u_n, and so on.

In the case $Ux = g + Ax$, $Sx = h + Bx$, where A and B are linear operators, system (11.3) may be written

$$\begin{aligned} u_n &= g + Au_n + Av_n, \\ v_n &= \sum_{i=0}^{k-1} B^i h + \sum_{i=1}^{k} B^i u_n + B^k v_{n-1}, \end{aligned} \tag{11.6}$$

so that

$$S_k\,(u,\, v) = \sum_{i=0}^{k-1} B^i h + \sum_{i=1}^{k} B^i u + B^k v.$$

It is easy to see that

$$\bar{R}_k = \left(I - \sum_{i=0}^{k} AB^i\right)^{-1}\left(g + \sum_{i=0}^{k-1} AB^i h + AB^k\right),$$

$$\bar{S}_k = \sum_{i=0}^{k-1} B^i h + \sum_{i=1}^{k} B^i\left(I - \sum_{j=0}^{k} AB^j\right)^{-1}\left(g + \sum_{l=0}^{k-1} AB^l h + AB^k\right) + B^k.$$

We can also establish convergence for algorithm (11.3) and derive appropriate error estimates without using explicit expressions for the resolvent operators, imposing conditions only on the original operators U and S.

Henceforth we shall assume that system (11.2) has at least one solution.

THEOREM 11.1. *Suppose that for any elements $x, y \in E$ the operators U and S satisfy the conditions*

$$\|Ux - Uy\| \leqslant L_U\,\|x - y\|, \tag{11.7}$$

$$\|Sx - Sy\| \leqslant L_S \|x - y\|, \tag{11.8}$$

where L_U and L_S are positive linear operators (Lipschitz operators). Suppose that the series $\Sigma_{i=0}^{\infty} \Delta_k^i a$ and $\Sigma_{i=0}^{\infty} \mathsf{E}_k^i a$, where

$$\Delta_k = \sum_{i=0}^{k} L_U L_S^i,$$

$$\mathsf{E}_k = \sum_{i=1}^{k} L_S^i (I - \Delta_k)^{-1} L_U L_S^k + L_S^k,$$

are uniformly convergent for any $a \in N$.

Then the systems of equations (11.3) *are uniquely solvable for u_n, v_n in E for any $v_0 \in E$, and the sequence $\{x_n\}$, $x_n = u_n + v_n$, converges to a solution of equation* (11.1) *for any $v_0 \in E$; moreover, the following error estimate holds*:

$$\|x - x_n\| \leqslant (I - L_U)^{-1} (I - \mathsf{E}_k)^{-1} \mathsf{E}_k^{n-p+1} \|\delta v_p\| \quad (1 \leqslant p \leqslant n), \tag{11.9}$$

where $\delta v_p = v_p - v_{p-1}$.

PROOF. That each of systems (11.3) is uniquely solvable follows from the fact that substitution of the expression for v_n from the second equation of the system into the first yields an equation for u_n in which the right-hand side is an operator satisfying a Lipschitz condition with Lipschitz operator Δ_k, where $\Sigma_{i=0}^{\infty} \Delta_k^i a$ is uniformly convergent for any $a \in N$.

Applying (11.7) and (11.8), we deduce from the first and second equations of (11.3) that

$$\|\delta u_n\| \leqslant L_U (\|\delta u_n\| + \|\delta v_n\|), \tag{11.10}$$

$$\|\delta v_n\| \leqslant \sum_{i=1}^{k} L_S^i \|\delta u_n\| + L_S^k \|\delta v_{n-1}\|. \tag{11.11}$$

Substituting the estimate for $\|\delta v_n\|$ from (11.11) into the right-hand side of (11.10), we obtain

$$\|\delta u_n\| \leqslant \Delta_k \|\delta u_n\| + L_U L_S^k \|\delta v_{n-1}\|,$$

whence

$$\|\delta u_n\| \leqslant (1 - \Delta_k)^{-1} L_U L_S^k \|\delta v_{n-1}\|. \tag{11.12}$$

Then, substituting the estimate (11.12) for $\|\delta u_n\|$ on the right of (11.11), we get

$$\|\delta v_n\| \leqslant \mathsf{E}_k \|\delta v_{n-1}\|. \tag{11.13}$$

Since $\Sigma_{i=0}^{\infty} \mathsf{E}_k^i a$ is uniformly convergent, it follows from (11.13) and (11.12) that $\{v_n\}$ and $\{u_n\}$ are Cauchy sequences, and so have limits $v \in E$ and $u \in E$.

We claim that u and v are solutions of system (11.5). Indeed,

$$\|u - U(u+v)\| \leqslant \|u - u_n\| + \|U(u_n + v_n) - U(u+v)\|$$
$$\leqslant \|u - u_n\| + L_U\|u_n - u\| + L_U\|v_n - v\|,$$
$$\|v - S_k(u, v)\| \leqslant \|v - v_n\| + \|S_k(u_n, v_{n-1}) - S_k(u, v)\|$$
$$\leqslant \|v - v_n\| + \sum_{i=1}^{k} L_S^i\|u_n - u\| + L_S^k\|v_{n-1} - v\|.$$

Hence, since $\|u_n - u\| \longrightarrow \theta$ and $\|v_n - v\| \longrightarrow \theta$ as $n \longrightarrow \infty$, it follows that $\|u - U(u+v)\| = \|v - S_k(u, v)\| = \theta$, so that u and v constitute a solution of system (11.5).

Uniqueness is readily proved by reductio ad absurdum. Suppose there are two solutions of system (11.5), say u, v and u', v'. Proceeding as in the derivation of (11.12) and (11.13), we see that

$$\|u - u'\| \leqslant (I - \Delta_k)^{-1} L_U L_S^k \|v - v'\|,$$
$$\|v - v'\| \leqslant E_k\|v - v'\|,$$

whence it follows that $v = v'$ and $u = u'$.

By assumption, system (11.2) has a solution and each of its solutions is a solution of (11.5); it follows, then, from the unique solvability of system (11.5) that the same is true of (11.2), and the solution of the latter is also that of (11.5). Thus the sequence $x_n = u_n + v_n$ converges to a solution of the original equation (11.1).

To establish the error estimate (11.9), we observe that by (11.10)

$$\|\delta u_n\| \leqslant (I - L_U)^{-1} L_U \|\delta v_n\|,$$

and since $\|\delta x_n\| \leqslant \|\delta u_n\| + \|\delta v_n\|$, we have

$$\|\delta x_n\| \leqslant [(I - L_U)^{-1} L_U + I]\,\|\delta v_n\| = (I - L_U)^{-1}\|\delta v_n\|.$$

We then use (11.13).

REMARK. We may replace Δ_k and E_k in Theorem 11.1 by

$$\bar{\Delta}_k = \sum_{i=1}^{k} L_S^i (I - L_U)^{-1} L_U, \qquad \bar{E}_k = (I - \bar{\Delta}_k)^{-1} L_S^k,$$

assuming them to satisfy the same conditions as Δ_k and E_k. This may be proved on the basis of the same inequalities (11.10) and (11.11).

We now turn to the case in which the assumptions of Theorem 11.1 hold not throughout the space E but only on some subset. Let D_1 and D_2 be any two closed subsets of E and D the set of all elements $x = u + v$, where $u \in D_1$ and $v \in D_2$. It is easily checked that if

$$Ux \in D_1, \quad Sy \in D_2 \tag{11.14}$$

for all $x, y \in D$, then $U(u+v) \in D_1$ and $S_i(u, v) \in D_2$ for $u \in D_1, v \in D_2$. Hence it follows that if (11.14) is true for arbitrary $x, y \in D$ and the assumptions of Theorem 11.1 hold for all the elements of D, then system (11.3) has a unique solution u_n, v_n for each $n, u_n \in D_1, v_n \in D_2$, provided only that $v_0 \in D_2$; moreover, the successive approximations $x_n = u_n + v_n \in D$ converge to the (unique) solution in D of equation (11.1), and the error estimate (11.9) is valid. The Lipschitz operators figuring in the error estimate must correspond to the set D. The proof of this proposition is entirely analogous to that of Theorem 11.1 with the sole difference that instead of E one considers the closed sets D_1, D_2 and D.

We now consider a few special cases in which (11.14) holds for $x, y \in D$.

I. Let D_1 and D_2 be balls with respective centers z_1 and z_2 and radii r_1 and r_2, and D the ball with center $z = z_1 + z_2$ and radius $r = r_1 + r_2$. Suppose that for all $x, y \in S(z, r)$ inequalities (11.7) and (11.8) hold. Then, if

$$\begin{aligned} \|Uz - z_1\| + L_U r &\leqslant r_1, \\ \|Sz - z_2\| + L_S r &\leqslant r_2, \end{aligned} \tag{11.15}$$

we have $Ux \in S(z_1, r_1)$ and $Sy \in S(z_2, r_2)$ for any $x, y \in D$. Indeed, if $x \in D$ and $y \in D$, i.e., $\|x - z\| \in r$ and $\|y - z\| \leqslant r$, then by (11.15)

$$\begin{aligned} \|Ux - z_1\| \leqslant \|Ux - Uz\| + \|Uz - z_1\| \leqslant \|Uz - z_1\| + L_U\|x - z\| \\ \leqslant \|Uz - z_1\| + L_U r \leqslant r_1, \\ \|Sy - z_2\| \leqslant \|Sy - Sz\| + \|Sz - z_2\| \leqslant \|Sz - z_2\| + L_S\|y - z\| \\ \leqslant \|Sz - z_2\| + L_S r \leqslant r_2, \end{aligned}$$

so that $Ux \in S(z_1, r_1)$ and $Sy \in S(z_2, r_2)$.

II. Suppose that for all $x, y \in S(z, r)$, where z is a fixed element of E and r will shortly be determined,

$$\|Ux - z_1\| \leqslant V_U(r), \quad \|Sy - z_2\| \leqslant V_S(r), \quad z = z_1 + z_2, \tag{11.16}$$

where V_U and V_S are operators from N to N.

If there exists an element $r^* \geqslant \theta$ such that

$$V_U(r) + V_S(r) \leqslant r, \tag{11.17}$$

then for any $x, y \in S(z, r^*)$

$$Ux \in S(z_1, V_U(r^*)), \quad Sy \in S(z_2, V_S(r^*)),$$

and the roles of the sets D, D_1 and D_2 may be filled by $S(z, r^*)$, $S(z_1, V_U(r^*))$ and $S(z_2, V_S(r^*))$, respectively. The proof is obvious. We need only observe that if $u \in D_1$ and $v \in D_2$, then $x = u + v \in D$, because

$$\|u + v - z\| \leqslant \|u - z_1\| + \|v - z_2\| \leqslant V_U(r^*) + V_S(r^*) \leqslant r^*.$$

III. Suppose that for any $x, y \in S(z, r)$, where z is a known element of E and r is to be determined

$$\|Ux - Uz\| \leqslant W_U(r)\|x - z\|, \tag{11.18}$$

$$\|Sy - Sz\| \leqslant W_S(r)\|y - z\|, \tag{11.19}$$

where $W_U(r)$ and $W_S(r)$ are operators on N.

If there exists $r^* \in N$ such that

$$\|Uz - z_1\| + \|Sz - z_2\| + W_U(r)\,r + W_S(r)\,r \leqslant r, \tag{11.20}$$

then for all $x, y \in S(z, r^*)$ we have

$$Ux \in S(z_1, r_1), \quad Sy \in S(z_2, r_2),$$

where

$$z_1 + z_2 = z, \quad r_1 = \|Uz - z_1\| + W_U(r^*)\,r^*, \quad r_2 = \|Sz - z_2\| + W_S(r^*)\,r^*,$$

and the element $u + v$, $u \in S(z_1, r_1)$, $v \in S(z_2, r_2)$, is in $S(z, r^*)$. In fact, applying (11.18) and (11.19) with $x, y \in S(z, r^*)$, we obtain

$$\|Ux - z_1\| \leqslant \|Ux - Uz\| + \|Uz - z_1\| \leqslant W_U(r^*)\|x - z\| + \|Uz - z_1\| \leqslant \|Uz - z_1\| + W_U(r^*)\,r^* = r_1,$$

$$\|Sy - z_2\| \leqslant \|Sy - Sz\| + \|Sz - z_2\| \leqslant W_S(r^*)\|y - z\| + \|Sz - z_2\| \leqslant \|Sz - z_2\| + W_S(r^*)\,r^* = r_2.$$

In addition, if $u \in S(z_1, r_1)$ and $v \in S(z_2, r_2)$, then $u + v \in S(z, r^*)$, since by assumption

$$\|u + v - z\| \leqslant \|u - z_1\| + \|v - z_2\| \leqslant \|Uz - z_1\| + W_U(r^*)\,r^* + \|Sz - z_2\| + W_S(r^*)\,r^* \leqslant r^*.$$

§12. Modified iterative projection method in some special cases

In this section we discuss convergence criteria and error estimates for algorithm (11.3) in the case that the operators U and S have a special form. The results complement those deduced in the preceding section under more general assumptions.

As before, we consider the operator equation (11.1), assuming that T is an operator in some complete space E lattice-normed by a K-lineal N. Let P be a projection operator ($P = P^2$), mapping E onto a subspace E_P, and set $Q = I - P$, where I is the identity operator. The first case we consider is (a) $U = PT$, $S = QT$. Clearly we then have $U + S = T$, and the system of operator equations (11.2) may be written

$$u = PT(u + v), \qquad v = QT(u + v). \tag{12.1}$$

The successive approximation algorithm (11.3) will be

$$u_n = PT(u_n + v_n), \qquad v_n = QS_k(u_n, v_{n-1}), \tag{12.2}$$

where the operators S_k are defined recursively by

$$S_1(u, v) = QT(u + v), \qquad S_i(u, v) = QT[u + S_{i-1}(u, v)], \tag{12.3}$$

where v_0 is an arbitrary element of the space $E - E_P = E_Q$.

Some convergence criteria and error estimates for algorithm (12.3) in a Banach space, in the special case $k = 1$, were considered by the present author [103].

Henceforth we shall assume that equation (12.1) has at least one solution.

It is readily seen that, for each n, solution of the system of operator equations (12.2) reduces to solution of operator equations in the subset E_P and evaluation of certain operators. Indeed, substituting the expression for v_n from the second equation of (12.2) into the first, we get the equation

$$u_n = PT[u_n + QS_k(u_n, v_{n-1})] \tag{12.4}$$

for u_n in E_P. To find v_n, we substitute u_n as determined from (12.4) into the second equation of (12.2) and evaluate the appropriate value of the operator.

As in the general case (algorithm (11.3)), this process of successive approximations may be analyzed by reducing it to a classical successive approximation procedure for some new equation, provided the requisite resolvent operators exist. The convergence criteria and error estimates thus found differ from those established in the general case, and may often prove more efficient. As is evident from (12.3), $u_n \in E_P$ and $v_n \in E_Q$, and so the operators $\bar{R}_k$ and $\bar{S}_k$ of §11 are defined by

$$\bar{R}_k = P\bar{R}_k Q, \quad \bar{S}_k = Q\bar{S}_k Q.$$

Thus v_n is defined by the recurrence relation

$$v_n = Q\bar{S}_k Q v_{n-1}. \tag{12.5}$$

One convergence criterion for the process (12.5) is that the series $\Sigma_{i=0}^{\infty} L^i_{Q\bar{S}_k Q}\, a$ is uniformly convergent to any $a \in N$ (the letter L with subscript denotes a Lipschitz operator for the appropriate operator). We then have the following error estimate:

$$\| x - x_n \| \leqslant L_{R_{PT}} \left(I - L_{Q\bar{S}_k Q}\right)^{-1} L^{n-p+1}_{Q\bar{S}_k Q} \| \delta v_p \| \qquad (1 \leqslant p \leqslant n), \tag{12.6}$$

where $x_n = u_n + v_n$, $\delta v_p = v_p - v_{p-1}$, and R_{PT} is the resolvent operator for

PT (i.e., an operator such that the solution of the equation $x = PTx + f$, if it exists, is given by $x = R_{PT}f$).

The proof is based on the observation that under these assumptions the sequence Qv_n defined by (12.5) is a Cauchy sequence and, since by (12.2) $x_n = PTx_n + Qv_n$, it follows that $x_n = T_{PT}Qv_n$, so that $\{x_n\}$ is also a Cauchy sequence. We then have

$$\|\delta x_n\| = \|x_n - x_{n-1}\| \leqslant L_{R_{PT}}\|Q\,\delta v_n\|.$$

Note that since in many cases $L_{QSQ} \leqslant L_S$, convergence criteria and error estimates involving the specific properties of P are indeed more efficient than those derived from general considerations.

Some convergence criteria and error estimates for algorithm (12.2) may be deduced from the general Theorem 11.1.

We shall establish some convergence criteria and error estimates on the assumption that inequality (10.1) is true for all $u, v \in E$.

THEOREM 2.1. *Suppose that inequality* (10.1) *is true for arbitrary $u, v \in E$, as well as the analogous inequalities for the operators PT and QT, with L_{TP} and L_{TQ} replaced by L_{PTP}, L_{PTQ} and L_{QTP}, L_{QTQ}. Assume that the series $\Sigma_{i=0}^{\infty} \Delta_k^i a$ and $\Sigma_{i=0}^{\infty} \mathrm{E}_k^i a$, where*

$$\Delta_k = L_{PTP} + L_{PTQ}\sum_{i=0}^{k-1} L_{QTQ}^i L_{QTP},$$

$$\mathrm{E}_k = \sum_{i=0}^{k-1} L_{QTQ}^i L_{QTP}(I - \Delta_k)^{-1} L_{PTQ} L_{QTQ}^k + L_{QTQ}^k,$$

are uniformly convergent for any $a \in N$.

Then u_n and v_n are uniquely determined from system (12.2) *for any $v_0 = Qz$ $(z \in E)$, and the successive approximations $x_n = u_n + v_n$ converge to a solution of equation* (11.1); *the following error estimate is valid*:

$$\|x - x_n\| \leqslant L_{R_{PT}}(I - \mathrm{E}_k)^{-1}\mathrm{E}_k^{n-p+1}\|\delta v_p\| \qquad (1 \leqslant p \leqslant n), \tag{12.7}$$

where $L_{R_{PT}}$ is a Lipschitz operator for the resolvent R_{PT} of PT.

PROOF. We first prove that system (12.2) is uniquely solvable for u_n and v_n. Substituting the expression for v_n from the second equation of (12.2) into the first and noting that $u_n = Pu_n$ and $v_n = Qv_n$, we obtain an equation for Pu_n:

$$Pu_n = PT[Pu_n + QS_k(Pu_n, Qv_{n-1})], \tag{12.8}$$

whose right-hand side satisfies a Lipschitz condition (see condition (10.1)) with an operator Δ_k such that $\Sigma_{i=0}^{\infty} \Delta_k^i a$ is uniformly convergent, by assumption. Thus equation (12.8) is uniquely solvable for Pu_n, which may then be substituted into the second equation of (12.2) to determine v_n. Thus x_n is uniquely determined.

From (12.8) it also follows that

$$\| P\delta u_n \| \leqslant \Delta_k \| P\delta u_n \| + L_{PTQ} L_{QTQ}^k \| Q\delta v_{n-1} \|,$$

whence

$$\| P\delta u_n \| \leqslant (I - \Delta_k)^{-1} L_{PTQ} L_{QTQ}^k \| Q\delta v_{n-1} \|. \tag{12.9}$$

By the second equation of (12.2),

$$\| Q\delta v_n \| \leqslant \sum_{i=0}^{k-1} L_{QTQ}^i L_{QTP} \| P\delta u_u \| + L_{QTQ}^k \| Q\delta v_{n-1} \|. \tag{12.10}$$

Substituting the estimate (12.9) for $\|P\delta u_n\|$ into the right-hand side of (12.10), we get

$$\| Q\delta v_n \| \leqslant E_k \| Q\delta v_{n-1} \|, \tag{12.11}$$

and this, by the uniform convergence of $\Sigma_{i=0}^{\infty} E_k^i a$, implies that $\{v_n\}$ is a Cauchy sequence, converging to a limit v. Since E is complete, $v \in E_Q$. Next, it follows from (12.9) that $\{u_n\}$ is also a Cauchy sequence, and therefore converges to a limit $u \in E_P$.

It is readily seen that u, v constitute a solution of the system

$$\begin{aligned} u &= PT(u + v), \\ v &= S_k(u, v). \end{aligned} \tag{12.12}$$

In fact, we have

$$\begin{aligned} \| u - PT(u + v) \| &\leqslant \| u - u_n \| + \| PT(u_n + v_n) - PT(u + v) \| \\ &\leqslant \| u - u_n \| + L_{PTP} \| u_n - u \| + L_{PTQ} \| v_n - v \|, \end{aligned}$$

$$\begin{aligned} \| v - S_k(u, v) \| &\leqslant \| v - v_n \| + \| S_k(u_n, v_{n-1}) - S_k(u, v) \| \\ &\leqslant \| v - v_n \| + \sum_{i=0}^{k-1} L_{QTQ}^i L_{QTP} \| u_n - u \| + L_{QTQ}^k \| v_{n-1} - v \|, \end{aligned}$$

and so, letting $n \to \infty$, we see that $\|u - PT(u + v)\| = \|v - S_k(u, v)\| = \theta$.

By the usual method, we now prove that the solution of (12.12) is unique. Indeed, the existence of two solutions u, v and u', v' would imply as before (see the proof of (12.9) and (12.11)) that

$$\begin{aligned} \| u - u' \| &\leqslant (I - \Delta_k)^{-1} L_{PTQ} L_{QTQ}^k \| v - v' \|, \\ \| v - v' \| &\leqslant E_k \| v - v' \|, \end{aligned}$$

whence it follows that $v = v'$ and $u = u'$.

Moreover, since every solution of system (12.1) is also a solution of system (12.12), and by assumption system (12.1) is solvable, it follows that the latter has a unique solution, coinciding with that of (12.12), and this solution is the limit of the successive approximations u_n, v_n.

The error estimate (12.7) is proved in the usual way. We need only observe that $x_n = R_{PT}v_n$.

REMARK 1. The operators Δ_k and E_k is the statement of Theorem 12.1 may be replaced respectively by

$$\overline{\Delta}_k = \sum_{i=0}^{k-1} L_{QTQ}^i L_{QTP} (I - L_{PTP})^{-1} L_{PTQ},$$
$$\overline{E}_k = (I - \overline{\Delta}_k)^{-1} L_{QTQ}^k,$$

assuming that they satisfy the same conditions as Δ_k and E_k.

This follows directly from the inequality

$$\| P\delta u_n \| \leqslant L_{PTP} \| P\delta u_n \| + L_{PTQ} \| Q\delta v_n \|, \tag{12.13}$$

which in turn follows from the first equation of (12.2) via (10.1).

REMARK 2. In the error estimate (12.7), we can replace the operator $L_{R_{PT}}$ by $(I - L_{PTP})^{-1}L_{PTQ} + I$. Indeed, since by (12.13)

$$\| P\delta u_n \| \leqslant (I - L_{PTP})^{-1} L_{PTQ} \| Q\delta v_n \|,$$

it follows that

$$\| \delta x_n \| \leqslant \| \delta u_n \| + \| \delta v_n \| \leqslant [(I - L_{PTP})^{-i} L_{PTQ} + I] \| \delta v_n \|.$$

We now consider the case that $Tx = f + T_1 x$, where $f \in E$ and T_1 has the property

$$T_1 (P + Q) = T_1 P + T_1 Q. \tag{12.14}$$

This condition is satisfied, for example, by linear operators, but also by some classes of nonlinear operators. Examples will be given in Chapter III.

When condition (12.14) is satisfied, the operators U and S figuring in algorithm (11.3) may be chosen as follows:

b) $U = TP$, $S = T_1 Q$;

c) $U = TP + PT_1 Q$, $S = QT_1 Q$.

In case (b), the system of operator equations (11.2) is clearly

$$\begin{aligned} u &= TP(u + v), \\ v &= T_1 Q (u + v), \end{aligned} \tag{12.15}$$

and algorithm (11.3) may be written

$$\bar{u}_n = TP(\bar{u}_n + \bar{v}_n),\quad \bar{v}_n = \bar{S}_k(\bar{u}_n, \bar{v}_{n-1}), \tag{12.16}$$

where the operators $\bar{S}_k$ are defined recursively by

$$\bar{S}_1(u, v) = T_1 Q(u + v),\quad \bar{S}_i(u, v) = T_1[Qu + Q\bar{S}_{i-1}(u, v)] \qquad (i = 2, 3, \ldots). \tag{12.17}$$

As in case (a), solution of system (12.16) reduces to solution of operator equations in the space E_P and evaluation of certain operators. Indeed, setting $\bar{x}_n = \bar{u}_n + \bar{v}_n$ and $z_n = Q\bar{u}_n + Q\bar{S}_{k-1}(\bar{u}_n, \bar{v}_{n-1})$, we use (12.17) and (12.14) to deduce from (12.16) the following operator equations for $P\bar{x}_n$ and z_n:

$$Px_n = PT(P\bar{x}_n + Qz_n),\quad Qz_n = QS_k(P\bar{x}_n, Qz_{n-1}), \tag{12.18}$$

and this system has the same form as the system (12.2) for u_n and v_n (if $v_0 = Qz_0$). The operators S_i are determined from (12.3). For $n = 1$, we have the system

$$P\bar{x}_1 = PT(P\bar{x}_1 + Qz_1),\quad Qz_1 = QS_{k-1}(P\bar{x}_1, QTP\bar{x}_1 + Q\bar{v}_0),\quad S_0(u, v) = v, \tag{12.19}$$

whose solution again reduces to solution of an operator equation in E_P and evaluation of certain operators. We note in addition thta $\bar{v}_n = T_1 z_n$, $\bar{u}_n = TP\bar{x}_n$ and $\bar{x}_n = T(P\bar{x}_n + z_n)$, so that $\bar{x}_n = Tx_n$, where x_n is the same element as in (12.2). Thus, investigation of algorithm (12.16) amounts to investigation of algorithm (12.2) under the assumptions of Theorem 12.1.

Since $\bar{x}_n = Tx_n$, the error estimate is given by

$$\|x - \bar{x}_n\| \leqslant L_T L_{R_{PT}} (I - \mathcal{E}_k)^{-1} \mathcal{E}_k^{n-p+1} \|\delta v_p\| \qquad (1 \leqslant p \leqslant n), \tag{12.20}$$

where $\mathcal{E}_k$ is defined in Theorem 12.1 and v_p is evaluated from system (12.2).

In addition, we have another estimate:

$$\|x - \bar{x}_n\| \leqslant L_T L_{R_{PT}} (I - \mathcal{E}_k)^{-1} \mathcal{E}_k^{n-p} (I - \bar{\Delta}_k)^{-1} L_{QTQ}^{k-1} \|Q\delta\bar{v}_p\| \qquad (1 \leqslant p \leqslant n). \tag{12.21}$$

Indeed, since the second equation of (12.18) may be written

$$Qz_n = QS_{k-1}(P\bar{x}_n, QTP\bar{x}_n + Q\bar{v}_{n-1}),$$

it follows that

$$\|Q\delta z_n\| \leqslant \sum_{i=0}^{k-1} L_{QTQ}^{i} L_{QTP} \|P\delta \bar{x}_n\| + L_{QTQ}^{k-1} \|Q\delta \bar{v}_{n-1}\|. \tag{12.22}$$

We now see from the first equation of (12.18) that

$$\|P\delta \bar{x}_n\| \leqslant L_{PTP} \|P\delta \bar{x}_n\| + L_{PTQ} \|Q\delta z_n\|,$$

whence

$$\|P\delta \bar{x}_n\| \leqslant (I - L_{PTP})^{-1} L_{PTQ} \|Q\delta z_n\|.$$

Substituting this into the right-hand side of (12.22), we obtain

$$\|Q\delta z_n\| \leqslant \bar{\Delta}_k \|Q\delta z_n\| + L_{QTQ}^{k-1} \|Q\delta \bar{v}_{n-1}\|,$$

whence

$$\|Q\delta z_n\| \leqslant (I - \bar{\Delta}_k)^{-1} L_{QTQ}^{k-1} \|Q\delta \bar{v}_{n-1}\|. \tag{12.23}$$

Using the equality $\|\delta v_p\| = \|Q\delta z_p\|$ and (12.23), we now obtain the required estimate (12.21) from (12.20).

We now consider case (c). The system of operator equations (11.3) is now

$$\tilde{u}_n = PT\left(\tilde{u}_n + \tilde{v}_n\right) + QTP\tilde{u}_n, \qquad \tilde{v}_n = Q\tilde{S}_k\left(\tilde{u}_n, \tilde{v}_{n-1}\right), \tag{12.24}$$

where the operators $\tilde{S}_k$ are defined recursively by

$$\begin{aligned} &\tilde{S}_1(u, v) = QT_1Q(u + v), \\ &\tilde{S}_i(u, v) = QT_1Q\left[u + \tilde{S}_{i-1}(u, v)\right] \qquad (i = 2, 3, \ldots, k). \end{aligned} \tag{12.25}$$

It is readily seen that in this case too solution of system (12.24) reduces to solution of operator equations in E_P and evaluation of the values of certain operators. In fact, since obviously

$$P\tilde{u}_n = PT\left(\tilde{u}_n + \tilde{v}_n\right), \tag{12.26}$$

we can substitute $Q\tilde{u}_n = QTP\tilde{u}_n$ into the second equation of (12.24), to obtain

$$\tilde{v}_n = Q\tilde{S}_k\left(QTP\tilde{u}_n, \tilde{v}_{n-1}\right), \tag{12.27}$$

and then, substituting this expression for $\tilde{v}_n$ and $Q\tilde{u}_n = QTP\tilde{u}_n$ into (12.26), we get an equation for $P\tilde{u}_n \in E_P$:

$$P\tilde{u}_n = PT\left[P\tilde{u}_n + QTP\tilde{u}_n + Q\tilde{S}_k\left(QTP\tilde{u}_n, \tilde{v}_{n-1}\right)\right]. \tag{12.28}$$

Solving this equation for $P\tilde{u}_n$, we then determine all the other unknowns, and finally $\tilde{x}_n = \tilde{u}_n + \tilde{v}_n$.

We now establish some convergence criteria and error estimates for algorithm (12.24).

THEOREM 12.2. *Suppose that* (10.1) *and the analogous inequality for the operators PT and QT are satisfied. Assume that the series* $\Sigma_{i=0}^{\infty} \widetilde{\Delta}_k^i\, a$ *and* $\Sigma_{i=0}^{\infty} \widetilde{\mathrm{E}}_k^i\, a$, *where*

$$\widetilde{\Delta}_k = L_{PTP} + L_{PTQ} \sum_{i=0}^{k} L_{QTQ}^i L_{QTP},$$

$$\widetilde{\mathrm{E}}_k = \sum_{i=1}^{k} L_{QTQ}^i L_{QTP} \left(I - \widetilde{\Delta}_k\right)^{-1} L_{PTQ} L_{QTQ}^k + L_{QTQ}^k,$$

are uniformly convergent for any $a \in N$.

Then $\widetilde{u}_n$ *and* $\widetilde{v}_n$ *are uniquely determined by system* (12.24) *for any* $\widetilde{v}_0 \in E$, *and the sequence* $\{\widetilde{x}_n\}, \widetilde{x}_n = \widetilde{u}_n + \widetilde{v}_n$, *converges to a solution of equation* (11.1) *with error estimate*

$$\left\| x - \widetilde{x}_n \right\| \leqslant M \left(I - \widetilde{\mathrm{E}}_k\right)^{-1} \widetilde{\mathrm{E}}_k^{n-p+1} \left\| \delta \widetilde{v}_p \right\| \qquad (1 \leqslant p \leqslant n), \tag{12.29}$$

where

$$M = L_{R_{PT}} \left[L_{QTP} \left(I - L_{PTP} - L_{PTQ} L_{QTP}\right)^{-1} L_{PTQ} + I \right].$$

PROOF. We first show that system (12.24) is uniquely solvable for $\widetilde{u}_n$ and $\widetilde{v}_n$. In fact, the operator on the right of (12.28) is readily seen (via condition (10.1)) to satisfy a Lipschitz condition with Lipschitz operator $\widetilde{\Delta}_k$ for which $\Sigma_{i=0}^{\infty} \widetilde{\Delta}_k^i\, a$ is uniformly convergent, by assumption. Consequently, we can treat (12.28) by the classical successive approximation technique, and it follows that $P\widetilde{u}_n$ is uniquely determined. Knowing $P\widetilde{u}_n$, we can determine $\widetilde{v}_n$ from (12.27) and then $\widetilde{u}_n$, evaluating the appropriate operators. Thus $\widetilde{u}_n$ and $\widetilde{v}_n$ are uniquely determined by (12.24).

We now prove that the successive approximations converge. By (12.28),

$$\left\| P\delta\widetilde{u}_n \right\| \leqslant \widetilde{\Delta}_k \left\| P\delta\widetilde{u}_n \right\| + L_{PTQ} L_{QTQ}^k \left\| \delta\widetilde{v}_{n-1} \right\|,$$

whence

$$\left\| P\delta\widetilde{u}_n \right\| \leqslant \left(I - \widetilde{\Delta}_k\right)^{-1} L_{PTQ} L_{QTQ}^k \left\| \delta\widetilde{v}_{n-1} \right\|, \tag{12.30}$$

and by (12.27)

$$\left\| \delta\widetilde{v}_n \right\| \leqslant \sum_{i=1}^{k} L_{QTQ}^i L_{QTP} \left\| P\delta\widetilde{u}_n \right\| + L_{QTQ}^k \left\| \delta\widetilde{v}_{n-1} \right\|.$$

Substituting (12.30) on the right of this inequality, we finally obtain

$$\left\|\delta\tilde{v}_n\right\| \leqslant \tilde{E}_k \left\|\delta\tilde{v}_{n-1}\right\|. \tag{12.31}$$

Thus $\{\tilde{v}_n\}$ is a Cauchy sequence, and it therefore has a limit $\tilde{v}$ which, since E is complete, lies in E. By (12.30), $\{P\tilde{u}_n\}$ is also a Cauchy sequence and therefore converges to a limit in E. Next, since $Q\tilde{u}_n = QTP\tilde{u}_n$, we have

$$\left\|\delta\tilde{u}_n\right\| \leqslant \left\|P\delta\tilde{u}_n\right\| + \left\|Q\delta\tilde{u}_n\right\| \leqslant (I + L_{QTP}) \left\|P\delta\tilde{u}_n\right\|, \tag{12.32}$$

and so $\{\tilde{u}_n\}$ is also a Cauchy sequence, converging to some limit $\tilde{u} \in E$.

We claim that $\tilde{u}$ and $\tilde{v}$ satisfy the system of equations

$$\begin{aligned} u &= PT(u+v) + QTPu, \\ v &= QS_k(u, v). \end{aligned} \tag{12.33}$$

Indeed,

$$\begin{aligned}
&\left\|\tilde{u} - PT\left(\tilde{u} + \tilde{v}\right) - QTP\tilde{u}\right\| \\
&\quad \leqslant \left\|\tilde{u} - \tilde{u}_n\right\| + \left\|PT\left(\tilde{u}_n + \tilde{v}_n\right) + QTP\tilde{u}_n - PT\left(\tilde{u} + \tilde{v}\right) - QTP\tilde{u}\right\| \\
&\quad \leqslant \left\|\tilde{u} - \tilde{u}_n\right\| + L_{PTP}\left\|P\tilde{u}_n - P\tilde{u}\right\| \\
&\qquad + L_{PTQ}L_{QTP}\left\|P\tilde{u}_n - P\tilde{u}\right\| + L_{PTQ}\left\|\tilde{v}_n - \tilde{v}\right\| + L_{QTP}\left\|P\tilde{u}_n - P\tilde{u}\right\|, \\
&\left\|\tilde{v} - Q\tilde{S}_k\left(\tilde{u}, \tilde{v}\right)\right\| \leqslant \left\|\tilde{v} - \tilde{v}_n\right\| + \left\|Q\tilde{S}_k\left(\tilde{u}_n, \tilde{v}_{n-1}\right) - Q\tilde{S}_k\left(\tilde{u}, \tilde{v}\right)\right\| \\
&\qquad \leqslant \sum_{i=1}^{k} L_{QTQ}^{i} L_{QTP}\left\|P\tilde{u}_n - P\tilde{u}\right\| + L_{QTQ}^{k}\left\|\tilde{v}_n - \tilde{v}\right\|,
\end{aligned}$$

and hence our assertion.

The uniqueness of the solution of (12.33) is proved in the usual manner. If there existed two solutions $\tilde{u}$, $\tilde{v}$ and $\tilde{u}'$, $\tilde{v}'$, the argument that led to (12.30), (12.31) and (12.32) would yield

$$\begin{aligned}
\left\|P\tilde{u} - P\tilde{u}'\right\| &\leqslant \left(I - \tilde{\Delta}_k\right)^{-1} L_{PTQ}L_{QTQ}^{k}\left\|\tilde{v} - \tilde{v}'\right\|, \\
\left\|\tilde{v} - \tilde{v}'\right\| &\leqslant \tilde{E}_k\left\|\tilde{v} - \tilde{v}'\right\|, \\
\left\|\tilde{u} - \tilde{u}'\right\| &\leqslant (I + L_{QTP})\left\|P\tilde{u} - P\tilde{u}'\right\|,
\end{aligned}$$

whence it follows that $\tilde{u} = \tilde{u}'$ and $\tilde{v} = \tilde{v}'$.

To establish the error estimate (12.29), we observe that by (12.24)

$$\tilde{x}_n = PT\tilde{x}_n + QTP\tilde{u}_n + \tilde{v}_n,$$

whence

$$\tilde{x}_n = R_{PT}\left(QTP\tilde{u}_n + \tilde{v}_n\right)$$

and

$$\left\|\delta\tilde{x}_n\right\| \leqslant L_{RPT}\left(L_{QTP}\left\|P\delta\tilde{u}_n\right\| + \left\|\delta\tilde{v}_n\right\|\right).$$

Substituting the estimate

$$\left\|P\delta\tilde{u}_n\right\| \leqslant \left(I - L_{PTP} - L_{PTQ}L_{QTP}\right)^{-1} L_{PTQ}\left\|\delta\tilde{v}_n\right\|,$$

which is easily deduced from (12.26), into the right-hand side of the above inequality, and using the fact that $Q\tilde{u}_n = QTP\tilde{u}_n$, we find that

$$\left\|\delta\tilde{x}_n\right\| \leqslant L_{RPT}\left[L_{QTP}\left(I - L_{PTP} - L_{PTQ}L_{QTP}\right)^{-1} L_{PTQ} + I\right]\left\|\delta\tilde{v}_n\right\|, \qquad (12.34)$$

whence it follows in particular that $\{\tilde{x}_n\}$ is also a Cauchy sequence. The error estimate (12.29) is now derived in the usual way from (12.34).

Assume that system (12.33) with $k = 1$ has a solution in E. Now, every solution of (12.33) with $k = 1$ is also a solution of the same system with $k > 1$. Thus it follows from the unique solvability of (12.33) for fixed k that it has a unique solution for $k = 1$. It is clear that $\tilde{x} = \tilde{u} + \tilde{v}$ is a solution of the original equation (11.1).

§13. Special case of the generalized iterative method for solution of certain types of operator equation

In this section we consider a variant of the generalized iterative method, applied to a certain type of operator equation, which generalizes a successive approximation technique suggested by Picone in [150] and [151]. The existence and uniqueness theorem proved in this section generalizes Picone's results for systems of integral equations.

Let E be a complete space normed by the elements of an archimedean lineal N (where N is a normed space). Consider the nonlinear operator equation

$$x = g + A(x)f(x), \qquad (13.1)$$

where $f(\cdot)$ is a (generally) nonlinear operator E; for each fixed $z \in E$, $A(z)$ is a linear operator on E; for each fixed $z \in E$, the expression $A(\cdot)f(z)$ defines a nonlinear operator on E, g is given element of E and x is the unknown.

We assume that the operator figuring in (13.1) satisfy the following conditions:

(1) $f(\cdot)$ satisfies a generalized Lipschitz condition:

$$\|f(x) - f(y)\| \leqslant L_f\|x - y\| \qquad (13.2)$$

for arbitrary $x, y \in E$.

(2) The operator $A(z)$, $z \in E$, has a Lipschitz operator L_A which is independent of z:

$$\| A(z)x \| \leqslant L_A \| x \| \qquad (x \in E). \tag{13.3}$$

(3) For every fixed $z \in E$, the operator $A(\cdot)f(z)$ has a Lipschitz operator depending on z:

$$\| A(x)f(z) - A(y)f(z) \| \leqslant L_A [\| f(z) \|] \| x - y \| \qquad (x, y \in E), \tag{13.4}$$

where $L_A [\| f(z) \|]$ is an operator such that if $a \leqslant b$ $(a, b \in N)$

$$L_A[a] \leqslant L_A[b]. \tag{13.5}$$

(4) The series

$$\sum_{i=0}^{\infty} K^i a, \quad \sum_{i=0}^{\infty} L^i a, \quad \sum_{i=0}^{\infty} M^i a,$$

where

$$K = L_A L_f, \quad L = L_f L_A, \quad M = (I - K)^{-1} L_A [(I - L)^{-1} \| f(g) \|],$$

are uniformly convergent for any $a \in N$.

THEOREM 13.1. *If conditions* (1)–(4) *are satisfied, then equation* (13.1) *has a unique solution x in E and the successive approximations x_n, uniquely determined by the equations*

$$x_n = g + A(x_{n-1}) f(x^n) \qquad (n = 1, 2, \ldots), \tag{13.6}$$

where x_0 is an arbitrary element of E, converge to x, with the error estimate

$$\| x - x_n \| \leqslant (I - M)^{-1} M^{n-p+1} \| \delta_p \| \qquad (1 \leqslant p \leqslant n,\ n = 1, 2, \ldots), \tag{13.7}$$

where $\delta_p = x_p - x_{p-1}$.

PROOF. We first observe that since by assumption the operator $A(z)f(\cdot)$ satisfies a Lipschitz condition for each $z \in E$, with an operator K such that $\Sigma_0^\infty K^i a$ is uniformly convergent for any $a \in N$, it follows that (13.6) is uniquely solvable for $x_n \in E$, for every $n = 1, 2, \ldots$ Next, we see from (13.6) that

$$\begin{aligned} \delta_n = x_n - x_{n-1} &= A(x_{n-1}) f(x_n) - A(x_{n-2}) f(x_{n-1}) \\ &= A(x_{n-1}) f(x_n) - A(x_{n-1}) f(x_{n-1}) + A(x_{n-1}) f(x_{n-1}) - A(x_{n-2}) f(x_{n-1}), \end{aligned}$$

and hence, by the assumptions of the theorem

$$\| \delta_n \| \leqslant L_A L_f \| \delta_n \| + L_A [\| f(x_{n-1}) \|] \| \delta_{n-1} \|$$

and

$$\| \delta_n \| \leqslant (I - K)^{-1} L_A [\| f(x_{n-1}) \|] \| \delta_{n-1} \|. \tag{13.8}$$

Again by (13.6), we have

$$f(x_n) = f[g + A(x_{n-1}) f(x_n)] = f[g + A(x_{n-1}) f(x_n)] - f(g) + f(g),$$

whence

$$\| f(x_n) \| \leqslant L_f \| A(x_{n-1}) f(x_n) \| + \| f(g) \| \leqslant L_f L_A \| f(x_n) \| + \| f(g) \|;$$

consequently,

$$\| f(x_n) \| \leqslant (I - L)^{-1} \| f(g) \| \qquad (n = 1, 2, \ldots). \tag{13.9}$$

Hence by (13.5) we obtain

$$L_A [\| f(x_{n-1}) \|] \leqslant L_A [(I - L)^{-1} \| f(g) \|],$$

so that inequality (13.8) may be sharpened:

$$\| \delta_n \| \leqslant (I - K)^{-1} L_A [(I - L)^{-1} \| f(g) \|] \| \delta_{n-1} \| = M \| \delta_{n-1} \|. \tag{13.10}$$

Since by assumption $\Sigma_0^\infty M^i a$ is uniformly convergent for any $a \in N$, it follows from (13.10) that $\{x_n\}$ is a Cauchy sequence and therefore has a limit x which, by the completeness of E, must lie in E.

We claim that x is a solution of (13.1). Indeed,

$$\begin{aligned} \| x - g - A(x) f(x) \| &\leqslant \| x - x_n \| + \| A(x_{n-1}) f(x_n) - A(x) f(x) \| \\ &\leqslant \| x - x_n \| + \| A(x_{n-1}) f(x_n) - A(x_{n-1}) f(x) \\ &\quad + A(x_{n-1}) f(x) - A(x) f(x) \| \leqslant \| x - x_n \| + L_A L_f \| x_n - x \| \\ &\quad + L_A [\| f(x) \|] \| x_{n-1} - x \|; \end{aligned}$$

whence we see, letting $n \longrightarrow \infty$, that $x - g - A(x) f(x) = \theta$.

The uniqueness of the solution is proved as usual. Assuming that besides x there is another solution x', we deduce as in the case of inequality (13.10) that

$$\| x - x' \| \leqslant M \| x - x' \|,$$

and hence, since $\Sigma_0^\infty M^i a$ is uniformly convergent for any $a \in N$, it follows that $x = x'$.

The error estimate is derived, as usual, from the inequality

$$\begin{aligned} \| x_{n+m} - x_n \| &\leqslant \| \delta_{n+m} \| + \cdots + \| \delta_{n+1} \| \\ &\leqslant (M^{m-1} + \cdots + I) \| \delta_{n+1} \| \leqslant \sum_{i=0}^{m-1} M^i M^{n-p+1} \| \delta_p \|, \end{aligned}$$

by letting $p \longrightarrow \infty$.

It is noteworthy that the above convergence criteria for algorithm (13.6) are sometimes satisfied in situations for which the convergence criteria for the classical successive approximation process

$$x_n = g + A(x_{n-1}) f(x_{n-1}),$$

fail to hold.

CHAPTER II

INVESTIGATION OF ITERATIVE PROJECTION METHODS FOR SOLUTION OF OPERATOR EQUATIONS IN BANACH SPACES

§14. Generalized iterative projection method in Banach spaces–general case.

As in Chapter I, we consider the (generally nonlinear) operator equation

$$x = Tx, \tag{14.1}$$

but the operator T will now have domain and range in a Banach space E.

A Banach space, as a complete normed space with numerical norm, is a special case of a space R metrized by elements of a set G and also of a space lattice-normed by elements of a K-lineal. Thus all the convergence criteria and error estimates for iterative projection methods established in Chapter I remain in force for operator equations in Banach spaces. However, we can use the specific properties of a Banach space (as a space with a numerical norm) and the specific properties of the various special cases and special projection operators, in order to derive many results that do not follow directly from the general theory of Chapter I. This applies, in particular, to various effective convergence criteria, some special iterative projection algorithms, and also many new and effective error estimates both for projection methods and for the more general iterative projection methods.

1. We first consider the generalized iterative method studied in §§5 and 8. Let $Tx = F(x, x)$, where $F(x, y) \in E$ for $x, y \in E$, so that equation (14.1) becomes

$$x = F(x, x). \tag{14.2}$$

By the general method, the successive approximations x_n to the solution of (14.2) are defined as solutions of operator equations

$$x_n = F_k(x_n, x_{n-1}) \qquad (n = 1, 2, \ldots), \tag{14.3}$$

where the operators F_k are defined recursively by

$$\begin{gathered} F_1(x, y) = F(x, y), \\ F_i(x, y) = F(x, F_{i-1}(x, y)) \qquad (i = 2, 3, \ldots), \end{gathered} \tag{14.4}$$

and x_0 is an arbitrary element of E.

Recall that in the general case the equation

$$x = F_k(x, x) \tag{14.5}$$

is not equivalent to equation (14.2), but any solution of (14.2) is also a solution of (14.5).

Suppose that for all x, y, z, t in some closed subset D of the Banach space E

$$\| F(x, y) - F(z, t) \| \leqslant q \| x - z \| + l \| y - t \|, \tag{14.6}$$

where q and l are constants.

The following theorem gives a simple convergence criterion and error estimate for algorithm (14.3).

THEOREM 14.1. *Suppose that equations* (14.2) *and* (14.3) *have solutions x and x_n, respectively, in the set D, and that condition* (14.6) *holds with*

$$q + l < 1. \tag{14.7}$$

Then for any $x_0 \in D$ the sequence $\{x_n\}$ of solutions $x_n \in D$ to equations (14.3) *converges to the unique solution in D of equation* (14.2), *and the following error estimate is valid*:

$$\| x - x_n \| \leqslant (1 - \varepsilon_k)^{-1} \varepsilon_k^{n-p+1} \| x_p - x_{p-1} \| \qquad (1 \leqslant p \leqslant n), \tag{14.8}$$

where

$$\varepsilon_k = l^k \left(1 - q \sum_{i=0}^{k-1} l^i \right)^{-1} < 1.$$

Moreover, $\epsilon_m < \epsilon_k$ for $m > k$, if $l \neq 0$.

This theorem is in fact a corollary of Theorem 8.1. It is sufficient to observe that by (14.7) $\delta_k = q \sum_{i=0}^{k-1} l^i < 1$ and $\epsilon_k < 1$. The proof of the last inequalities is elementary. We have

$$\delta_k < q \sum_{i=0}^{\infty} l^i = \frac{q}{1-l} < 1$$

and for any $k = 1, 2, \ldots$

$$\frac{l^{k+1}}{1 - q \sum_{i=0}^{k} l^i} < \frac{l^k}{1 - q \sum_{i=0}^{k-1} l^i},$$

since obviously

$$l \left(1 - q \sum_{i=0}^{k-1} l^i \right) < 1 - q \sum_{i=0}^{k} l^i.$$

But since $\epsilon_1 = l/(1 - q) < 1$, it follows that $\epsilon_k < 1$ and $\epsilon_m < \epsilon_k$ for $m > k$.

We shall give another convergence criterion for algorithm (14.3) and a corresponding error estimate, which are often less restrictive than those embodied in Theorem 14.1.

Suppose that for any $v \in D$ the operator $F(\cdot, v)$ has a resolvent operator R—an operator such that the unique solution of the equation $x = F(x, v)$ (assuming that it exists) is given by $x = Rv$, $x \in D$. In this case we have

THEOREM 14.2. *Suppose that the operator R satisfies a Lipschitz condition for the elements of D, with a constant $l_R < 1$, and*

$$ql_R + l < 1, \tag{14.9}$$

where the constants q and l are as in Theorem 14.1, and equations (14.3) have solutions $x_n \in D$ for any $x_0 \in D$.

Then $x_n \in D$ are uniquely determined by equations (14.3) for any $x_0 \in D$ and the sequence $\{x_n\}$ converges to the unique solution $x \in D$ of equation (14.2), with the error estimate

$$\|x - x_n\| \leqslant (1 - \varepsilon_k)^{-1} \varepsilon_k^{n-p} \|x_p - x_{p-1}\| \quad (1 \leqslant p \leqslant n), \tag{14.10}$$

where

$$\varepsilon_k = l_R l^{k-1} \left(1 - l_R q \sum_{i=0}^{k-2} l^i\right)^{-1} < 1.$$

Here $\epsilon_m < \epsilon_k$ for $m > k$, if $l \neq 0$ and $l_R \neq 0$.

PROOF. By the recurrence relations (14.4), we can write (14.3) as

$$x_n = F[x_n, F_{k-1}(x_n, x_{n-1})] \qquad (F_0(x, y) = y),$$

whence

$$x_n = RF_{k-1}(x_n, x_{n-1}) \qquad (n = 1, 2, \ldots). \tag{14.11}$$

Clearly, the operator on the right of (14.11) satisfies the condition

$$\|RF_{k-1}(x, y) - RF_{k-1}(z, t)\| \leqslant l_R q \sum_{i=0}^{k-2} l^i \|x - z\| + l_R l^{k-1} \|y - t\|$$

and, as is readily seen, $\delta_k = l_R q \Sigma_0^{k-2} l^i < 1, \epsilon_k < 1$, and $\epsilon_m < \epsilon_k$ for $m > k$. The rest of the proof is analogous to that of Theorem 8.1.

2. In §9 we considered three special cases of the generalized iterative method (14.3), corresponding to three different structures of the operator $F(x, y)$ in terms of the given operator T:

a) $F(x, y) = PTx + QTy$,

b) $F(x, y) = T(Px + Qy)$,

c) $F(x, y) = PTx + QT(Px + Qy)$,

where P is a projection operator mapping E onto a finite- or infinite-dimensional subspace E_P. We showed there that case (b) may be reduced to case (a).

Investigation of algorithms (14.3) in cases (a) and (b), for the operator equation (14.1) in a Banach space, may be based directly on the results of Theorem 9.1, as well as on subsection 1 of the present section. We must then set $q = l_{PT}$ and $l = l_{QT}$, where the l's denote Lipschitz constants for the appropriate operators.

We now turn to case (c), assuming that the operators PT and QT satisfy Lipschitz conditions with constants l_{PT} and l_{QT}, respectively.

THEOREM 14.3. *If*

$$l_{PT} + l_{QT} < 1, \tag{14.12}$$

then in case (c) *equation* (14.3) *is uniquely solvable for* x_n *in* E, *and the sequence of solutions* $\{x_n\}$ *converges to the unique solution of equation* (14.1) *in* E, *with error estimate*

$$\|x - x_n\| \leqslant (1 - l_{PT})^{-1} (1 - \varepsilon_k)^{-1} \varepsilon_k^{n-p+1} \| Q(x_p - x_{p-1}) \| \quad (1 \leqslant p \leqslant n), \tag{14.13}$$

where

$$\varepsilon_k = l_{QT}^k \left[1 - l_{PT} (1 - l_{PT})^{-1} \sum_{i=1}^{k} l_{QT}^i \right]^{-1} < 1,$$

and $\epsilon_m < \epsilon_k$ *for* $m > k$, *if* $l_{QT} \neq 0$.

PROOF. That the equations for x_n are uniquely solvable follows from the fact that the operators on their right satisfy a Lipschitz condition with constant $\delta_k = l_{PT} \Sigma_0^k l_{QT}^i$; by (14.12) this constant is less than 1, since $\delta_k \leqslant l_{PT}/(1 - l_{QT}) < 1$. Next, it is easy to see that $\epsilon_{k+1} < \epsilon_k$ for any $k = 1, 2, \ldots$, since by (14.12)

$$l_{QT} \left[1 - l_{PT} (1 - l_{PT})^{-1} \sum_{i=1}^{k-1} l_{QT}^i \right] < 1 - l_{PT} (1 - l_{PT})^{-1} \sum_{i=1}^{k} l_{QT}^i;$$

and since, as is readily deduced from (14.12),

$$\varepsilon_1 = l_{QT} \left[1 - l_{PT} (1 - l_{PT})^{-1} l_{QT} \right]^{-1} < 1,$$

it follows that $\epsilon_k < 1$ for any k and $\epsilon_m < \epsilon_k$ for $m > k$.

The unique solvability of (14.1) is proved as follows. By (14.12), the operator T satisfies a Lipschitz condition with constant less than 1, since

$$\begin{aligned} \| Tx - Ty \| &= \| PTx + QTx - PTy - QTy \| \\ &\leqslant \| PTx - PTy \| + \| QTx - QTy \| \\ &\leqslant l_{PT} \| x - y \| + l_{QT} \| x - y \| = (l_{PT} + l_{QT}) \| x - y \|. \end{aligned}$$

The rest of the theorem is an immediate corollary of Theorem 9.2.

The results of §9 also yield convergence criteria and error estimates for iterative projection methods in the case that the conditions hold not throughout E but only in some subset.

3. Finally, we consider the case that for arbitrary $x, y, \bar{x}, \bar{y}$,

$$\|T(Px + Qy) - T(P\bar{x} + Q\bar{y})\| \leqslant q_{TP}\|(P(x - \bar{x})\| + q_{TQ}\|Q(y - \bar{y})\|, \quad (14.14)$$

where q_{TP} and q_{TQ} are constants, and analogous inequalities hold for the operators PT and QT, with q_{TP} and q_{TQ} replaced respectively by q_{PTP}, q_{PTQ} and q_{QTP}, q_{QTQ}.

THEOREM 14.4. *If $q_{PTP} < 1$ and*

$$q_{PTP} + q_{PTQ}q_{QTP} + q_{QTQ}(1 - q_{PTP}) < 1, \quad (14.15)$$

then the successive approximations x_n defined by (14.3) *in cases* (a) *and* (b) *are unique for any k and $x_0 \in E$, and converge to a solution of equation* (14.1); *error estimates are obtained from* (10.2) *and* (10.3) *by the substitutions*

$$L_{PTP} = q_{PTP}, \quad L_{PTQ} = q_{PTQ}, \quad L_{QTP} = q_{QTP}, \quad L_{QTQ} = q_{QTQ}.$$

Moreover, $\epsilon_m < \epsilon_k$ for $m > k$, if $q_{QTQ} \neq 0$.

To derive this theorem from Theorem 10.1, it suffices to show that the inequalities $q_{PTP} < 1$ and (14.15) imply the estimates

$$\delta_k = q_{PTP} + q_{PTQ}q_{QTP}\sum_{i=0}^{k-2} q_{QTQ}^{i} < 1, \quad \varepsilon_k = q_{QTQ}^{k} + \frac{q_{QTP}q_{PTQ}\sum_{i=0}^{k-1} q_{QTQ}^{i+k-1}}{1 - \delta_k} < 1.$$

But these inequalities are indeed true, for by the assumption $q_{PTP} < 1$ and (14.15) we have the inequalities $q_{QTQ} < 1$ and

$$\delta_k < q_{PTP} + \frac{q_{PTQ}q_{QTP}}{1 - q_{QTQ}} < 1.$$

Finally, it is not difficult to show that $\epsilon_{k+1} < \epsilon_k$ and

$$\varepsilon_1 = q_{QTQ} + \frac{q_{QTP}q_{PTQ}}{1 - q_{PTP}} < 1.$$

THEOREM 14.5. *If $q_{PTP} < 1$ and inequality* (14.15) *holds, then the successive approximations x_n are uniquely determined by equations* (14.3) *in case* (c) *for any k and $x_0 \in E$, and they converge to a solution of equation* (14.1); *as error estimate, we have* (10.16), *with the L's replaced by the appropriate q's.*

Moreover, $\epsilon_m < \epsilon_k$ for $m > k$ if $q_{QTQ} \neq 0$.

This theorem is easily deduced from Theorem 10.2, for by (14.15)

$$\delta_k = q_{PTP} + q_{PTQ} q_{QTP} \sum_{i=0}^{k-1} q_{QTQ}^i < 1, \quad \varepsilon_k = q_{QTQ}^k + \frac{q_{QTP} q_{PTQ} \sum_{i=0}^{k-1} q_{QTQ}^{i+k}}{1 - \delta_k} < 1.$$

These inequalities, and the inequalities $\epsilon_m < \epsilon_k$ $(m > k)$, are easily proved as in Theorem 14.4.

§15. Modified generalized iterative projection method in a Banach space—general case

The subject of this section is a modified generalized iterative projection method for solution of operator equations in Banach space, studied in §§11 and 12 for the general case of lattice-normed spaces.

1. As before, we consider the operator equation (14.1). Suppose that the operator T is the sum of two operators U and S, $T = U + S$, where U and S are operators on a Banach space E. Recall that according to this method (14.1) is replaced by a system of operator equations

$$u = U(u + v), \qquad v = S(u + v), \tag{15.1}$$

and the successive approximations to the solution of this system are the solutions of the systems

$$u_n = U(u_n + v_n), \qquad v_n = S_k(u_n, v_{n-1}), \; v_0 \in E. \tag{15.2}$$

The operators S_k are defined recursively by

$$\begin{aligned} &S_1(u, v) = S(u + v), \\ &S_i(u, v) = S[u + S_{i-1}(u, v)] \qquad (i = 2, 3, \ldots). \end{aligned} \tag{15.3}$$

The successive approximations to the solution of (14.1) are defined as $x_n = u_n + v_n$.

The algorithm (15.2) was investigated in §11 under quite general assumptions concerning U and S. The following theorem is an analog of Theorem 11.1.

THEOREM 15.1. *Let the operators U and S satisfy Lipschitz conditions with constants l_U and l_S such that*

$$l_U + l_S < 1. \tag{15.4}$$

Then system (15.2) is uniquely solvable for u_n and v_n in E for and $v_0 \in E$, and the sequence $\{x_n\}, x_n = u_n + v_n$, converges to the unique solution of equation (14.1), with error estimate

$$\|x - x_n\| \leqslant (1 - l_U)^{-1} (1 - \varepsilon_k)^{-1} \varepsilon_k^{n-p+1} \|\delta v_{p-1}\| \qquad (1 \leqslant p \leqslant n), \tag{15.5}$$

where

$$\varepsilon_k = l_S^k + \left(1 - l_U \sum_{i=0}^{k} l_S^i\right)^{-1} l_U \sum_{i=1}^{k} l_S^{i+k} < 1, \quad \delta v_p = v_p - v_{p-1}.$$

Moreover, $\epsilon_m < \epsilon_k$ for $m > k$.

For the proof it is sufficient to verify that by (15.4) $\delta_k = l_U \sum_{i=0}^{k} l_S^i < 1$, $\epsilon_{k+1} < \epsilon_k$ and $\epsilon_1 < 1$. The rest of the theorem follows directly from Theorem 11.1.

2. Suppose that for all $x, y, \bar{x}, \bar{y} \in E$ condition (14.14) holds. We consider the cases

a) $U = PT, \quad S = QT,$

b) $U = TP, \quad S = T_1Q \qquad (Tx = f + T_1x, \quad f \in E),$

where P is a projection of E onto a subspace E_P. As shown in §12, solution of (15.2) in these cases reduces to solution of a certain operator equation in E_P and evaluation of certain operators. The convergence of the successive approximations defined by (15.2) may be investigated using the results of §12.

A simple convergence criterion for algorithm (15.2) is provided by the following result.

THEOREM 15.2. *If $q_{PTP} < 1$ and inequality* (14.15) *holds, then the system of operator equations* (15.2)*(cases* (a) *and* (b)) *is uniquely solvable and the sequences* $\{x_n\}$ *and* $\{\bar{x}_n\}$, $x_n = u_n + v_n$, $\bar{x}_n = \bar{u}_n + \bar{v}_n$ *(where u_n, v_n are solutions of system* (15.2) *in case* (a) *and $\bar{u}_n$, $\bar{v}_n$ solutions of* (15.2) *in case* (b)) *converge to a solution of equation* (14.1) *for any $v_0 \in E$ and $\bar{v}_0 = Qz$ $(z \in E)$, and the following error estimates hold:*

In case (a):

$$\|x - x_n\| \leqslant l_{RPT}(1 - \varepsilon_k)^{-1} \varepsilon_k^{n-p+1} \|\delta v_p\| \qquad (1 \leqslant p \leqslant n); \quad (15.6)$$

In case (b):

$$\|x - \bar{x}_n\| \leqslant l_T l_{RPT}(1 - \varepsilon_k)^{-1} \varepsilon_k^{n-p+1} \|\delta v_p\| \qquad (1 \leqslant p \leqslant n), \quad (15.7)$$

$$\|x - \bar{x}_n\| \leqslant l_T l_{RPT}(1 - \varepsilon_k)^{-1} \varepsilon_k^{n-p} (1 - \bar{\delta}_k)^{-1} l_{QTQ}^{k-1} \|Q\delta \bar{v}_p\| \qquad (15.8)$$

$$(1 \leqslant p \leqslant n),$$

where, as before, the l's denote Lipschitz constants for the appropriate operators, and ϵ_k and $\bar{\delta}_k$ are given by

$$\varepsilon_k = \frac{q_{QTP} q_{PTQ} \sum_{i=0}^{k-1} q_{QTQ}^{i+k}}{1 - q_{PTP} - q_{PTQ} q_{QTP} \sum_{i=0}^{k-1} q_{QTQ}^{i}} + q_{QTQ}^k = \frac{q_{QTQ}^k}{1 - \bar{\delta}_k} < 1,$$

$$\bar{\delta}_k = \frac{q_{QTP} q_{PTQ}}{1 - q_{PTP}} \sum_{i=0}^{k-1} q_{QTQ}^i.$$

Moreover, $\epsilon_m < \epsilon_k$ for $m > k$, if $q_{QTQ} \neq 0$.

The main part of the proof rests on the results of §12. We need only observe that the inequalities $q_{PTP} < 1$ and (14.15) imply that

$$\delta_k = q_{PTP} + q_{PTQ}q_{QTP}\sum_{i=0}^{k-1} q^i_{QTQ} < 1, \ \bar{\delta}_k < 1, \ \varepsilon_{k+1} < \varepsilon_k \qquad \varepsilon_1 < 1.$$

We now consider the case

c) $U = PT + QTP, \ S = QT_1Q \qquad (x = f + T_1x, \ f \in E)$.

As shown in §12, in this case solution of system (15.2) again reduces to solution of certain new operator equations in E_P and evaluation of certain operators. Information about the convergence of algorithm (15.2) in this case is furnished by the following result.

THEOREM 15.3. *Suppose that inequality* (14.14) *is true for all* $x, y, \bar{x}, \bar{y} \in E$. *If* $q_{PTP} < 1$ *and inequality* (14.15) *is valid, then* u_n *and* v_n *are uniquely determined by* (15.2) *for any* $v_0 \in E$ *and the sequence* $\{x_n\}, x_n = u_n + v_n$, *converges to a solution of equation* (14.1), *with error estimate*

$$\| x - x_n \| \leqslant M (1 - \varepsilon_k)^{-1} \varepsilon_k^{n-p+1} \| \delta v_p \| \qquad (1 \leqslant p \leqslant n), \qquad (15.9)$$

where

$$M = l_{R_{PT}} \left[q_{QTP} q_{PTQ} \left(1 - q_{PTP} - q_{PTQ} q_{QTP}\right)^{-1} + 1 \right],$$

$$\varepsilon_k = \frac{q_{QTP} q_{PTQ} \sum_{i=1}^{k} q_{QTQ}^{i+k}}{1 - \delta_k} + q^k_{QTQ}, \qquad \delta_k = q_{PTP} + q_{PTQ} q_{QTP} \sum_{i=0}^{k} q^i_{QTQ}.$$

To prove the theorem, we need only apply Theorem 12.2, noting that the inequalities $q_{PTP} < 1$ and (14.15) imply $\delta_k < 1, \epsilon_{k+1} < \epsilon_k$ and $\epsilon_1 < 1$.

§16. Hilbert space and orthogonal projections. Error estimate for projection method

1. Hilbert spaces constitute an important special class of Banach spaces and of the more general lattice-normed spaces.

An abstract Hilbert space H is a set of elements $x, y, z, \ldots$ with the following properties:

1) H is a linear space.

2) The norm of an element x in H is defined by

$$\| x \| = \sqrt{(x, x)},$$

where (x, y) denotes the scalar product of $x, y \in H$—a complex (or real) number satisfying the following axioms:

a) $(y, x) = \overline{(x, y)}$.

b) $(\lambda x + \mu y, z) = \lambda(x, z) + \mu(y, z)$, where λ and μ are complex numbers.

c) $(x, x) \geqslant 0$, and $(x, x) = 0$ if and only if $x = 0$.

The main type of convergence in a Hilbert space is convergence in norm, or *strong convergence*.

Examples of Hilbert spaces are the space l_2 of number sequences $x = (x_1, x_2, \ldots)$ with scalar product

$$(x, y) = \sum_{i=1}^{\infty} x_i \overline{y_i} \qquad (x = \{x_i\}, \quad y = \{y_i\}),$$

the space L_2 of square-summable functions on an interval [a, b], with scalar product

$$(x, y) = \int_a^b x(t)\, y(t)\, dt \qquad (x(t),\ y(t) \in L_2),$$

and others.

An important type of operator in a Hilbert space is the orthogonal projection. An orthogonal projection is a projection ($P = P^2$) which maps the space H onto a subspace H_P such that for any $x \in H$ the elements Px and Qx, where $Q = I - P$ (I is the identity operator) are orthogonal, i.e., $(Px, Qy) = 0$.

A linear operator P in H is an orthogonal projection if and only if 1) P is selfadjoint, i.e., $(Px, y) = (x, Py)$ for any $x, y \in H$, and 2) $P(Px) = Px$ for any $x \in H$.

An important property of an orthogonal projection is that for any $x, y \in H$

$$\| Px + Qy \|^2 = \| Px \|^2 + \| Qy \|^2. \tag{16.1}$$

Indeed, since P and $Q = I - P$ are selfadjoint and $PQ = QP = 0$,

$$\begin{aligned} \| Px + Qy \|^2 &= (Px + Qy,\ Px + Qy) = (Px, Px) + (Px, Qy) \\ &+ (Qy, Px) + (Qy, Qy) = (Px, Px) + (Qy, Qy) = \| Px \|^2 + \| Qy \|^2. \end{aligned}$$

Property (16.1) may frequently be used to sharpen error estimates in projection methods, and also to weaken conditions for convergence and to sharpen error estimates in various iterative projection methods. The estimates derived in this way do not follow from the general theory of Chapter I.

2. We shall establish a few simple error estimates in the projection method for solution of operator equations (14.1) where T is an operator in a Hilbert space H. We shall see that the use of (16.1) often yields sharper estimates than those following directly from the general comparison theorems [169].

The main point of the projection method [78] for solution of the operator equation

$$x = Tx \tag{16.2}$$

is to replace the equation by an "approximating" equation

$$y = PTy \tag{16.3}$$

in the subspace E_P onto which the operator P maps the original space E, and to take the solution of equation (16.3) as an approximation to that of equation (16.2).

Let us consider the case in which the projection is orthogonal. We have the following result.

THEOREM 16.1. *Suppose that equation* (16.2) *has a solution x and equation* (16.3) *a solution in some convex subset D of a Hilbert space H. Let the operators PT and QT, $Q = I - P$, satisfy Lipschitz conditions*:

$$\| PTx - PTy \| \leqslant q_{PT} \| x - y \|, \tag{16.4}$$

$$\| QTx - QTy \| \leqslant q_{QT} \| x - y \| \tag{16.5}$$

for any $x, y \in D$. If

$$q_{PT}^2 + q_{QT}^2 < 1, \tag{16.6}$$

then the error $\|x - y\|$ of the approximate solution satisfies

$$\| x - y \| \leqslant \left(\sqrt{1 - q_{PT}^2} - q_{QT} \right)^{-1} \| QTy \|. \tag{16.7}$$

PROOF. Since x and y are solutions of equations (16.2) and (16.3), respectively, it follows that

$$x - y = Tx - PTy = Tx - PTy + PTx - PTx = QTx + (PTx - PTy).$$

Hence, by property (16.1), which is valid here because P is an orthogonal projection, and by condition (16.4), we have

$$\| x - y \|^2 = \| QTx \|^2 + \| PTx - PTy \|^2 \leqslant \| QTx \|^2 + q_{PT}^2 \| x - y \|^2.$$

Consequently,

$$\| x - y \| \leqslant (1 - q_{PT}^2)^{-\frac{1}{2}} \| QTx \|. \tag{16.8}$$

Furthermore, by (16.6), $q_{QT}(1 - q_{PT}^2)^{-1/2} < 1$, and thus, by (16.8) and (16.5),

$$\| x - y \| \leqslant (1 - q_{PT}^2)^{-\frac{1}{2}} (\| QTx - QTy \| + \| QTy \|)$$

$$\leqslant (1 - q_{PT}^2)^{-\frac{1}{2}} (q_{QT} \| x - y \| + \| QTy \|).$$

Solving this inequality, we obtain (16.7).

Let us assume now that T satisfies the assumptions of the contraction mapping principle in a ball $S(Pg, r)$ of radius r centered at Pg, where g is some element of H. In other words, $Tx \in S(Pg, r)$ for $x \in S(Pg, r)$ and $q_T < 1$ (where q's as usual, denote Lipschitz constants for the appropriate operators). We know that equation (16.2) will then have a unique solution in $S(Pg, r)$. On the other hand, if we know only that $Tx \in S(Pg, r)$ for $x \in S(Pg, r)$ and T is a compact operator, then the Schauder fixed-point principle guarantees the existence of a (not necessarily unique) solution of equation (16.2) in $S(Pg, r)$. Using the property

$$\|Px\| \leqslant \|x\| \qquad (x \in H), \tag{16.9}$$

which follows in an obvious manner from (16.1), one easily shows that in this case equation (16.3) again has a solution in $S(Pg, r)$.

Indeed, by (16.9), for any $x \in S(Pg, r)$ we have

$$\|PTx - Pg\| \leqslant \|Tx - Pg\| \leqslant r,$$

i.e., the compact operator PT maps $S(Pg, r)$ into itself. Consequently, by the Schauder principle, equation (16.3) has a solution $\bar{y}$ in $S(Pg, r)$. If moreover the operator T satisfies a Lipschitz condition in $S(Pg, r)$ with constant $q_T < 1$, this solution is unique in $S(Pg, r)$. Indeed, for any $x, y \in S(Pg, r)$,

$$\|PTx - PTy\| \leqslant \|Tx - Ty\| \leqslant q_T \|x - y\|,$$

and so $q_{PT} \leqslant q_T < 1$. Thus the operator PT satisfies the conditions of the contraction mapping principle.

It should be emphasized that in actual solution of various types of operator equations the exact Lipschitz constants are most frequently not available, and one has only certain estimates. If the operators in question are defined in Hilbert spaces, the Lipschitz constants q_T, q_{PT} and q_{QT} may often be estimated through quantities $\bar{q}_T$, $\bar{q}_{PT}$ and $\bar{q}_{QT}$ such that

$$\bar{q}_T^2 = \bar{q}_{PT}^2 + \bar{q}_{QT}^2. \tag{16.10}$$

Estimates of this kind will be considered in Chapter III in connection with iterative projection methods as applied to special classes of operator equations. Their appearance is bound up with the use of integral inequalities and inequalities for sums. When condition (16.10) is satisfied, the error estimate (16.7) and the error estimate obtained when q_{PT} and q_{QT} in (16.7) are replaced by $\bar{q}_{PT}$ and $\bar{q}_{QT}$ are generally sharper than the estimate

$$\|x - y\| \leqslant (1 - \bar{q}_T)^{-1} \|QTy\|,$$

which follows directly from the comparison theorem [169].

To prove this, it will suffice to show that

$$\sqrt{1-\bar{q}_{PT}^2}-\bar{q}_{QT}\geqslant 1-\bar{q}_T. \tag{16.11}$$

Supposing the contrary, we obtain

$$1-\bar{q}_T+\bar{q}_{QT}\geqslant\sqrt{1-\bar{q}_{PT}^2},$$

whence, squaring both (positive!) sides of this inequality, we get

$$1+\bar{q}_T^2+\bar{q}_{QT}^2-2\bar{q}_T+2\bar{q}_{QT}-2\bar{q}_T\bar{q}_{QT}\geqslant 1-\bar{q}_{PT}^2.$$

In view of (16.10), this gives

$$\bar{q}_T^2\geqslant\bar{q}_T-\bar{q}_{QT}(1-\bar{q}_T)\quad\text{or}\quad(\bar{q}_T-\bar{q}_{QT})(1-\bar{q}_T)\leqslant 0,$$

and if $q_T<1$ this is possible only if $\bar{q}_T=\bar{q}_{QT}$, because $\bar{q}_{QT}\leqslant\bar{q}_T$. We have

$$\sqrt{1-\bar{q}_{PT}^2}-\bar{q}_{QT}=1-\bar{q}_T,$$

so that $\bar{q}_{PT}=0$. This proves (16.11).

REMARK. The error estimate (16.7) may be sharpened by the same reasoning as at the end of §2, or by considering nonlinear majorants.

§17. Investigation of generalized iterative projection methods for operator equations in Hilbert space

In this section we consider the general iterative projection methods which were studied for equations in lattice-normed spaces in §9 and in Banach spaces in §14. We shall assume that the operator T in equation (16.2) ($x = Tx$) is defined on a Hilbert space H.

Let P be an orthogonal projection of E onto a finite- or infinite-dimensional subspace E_P. As usual, Q will denote the difference $I - P$.

We shall consider three algorithms, corresponding to three different choices of the operator $F(x, y)$ in the general iterative method (14.3):

a) $F(x,y)=PTx+QTy$;

b) $F(x,y)=T(Px+Qy)$;

c) $F(x,y)=PTx+QT(Px+Qy)$.

In case (a) the successive approximations u_n to the solution of (16.2) are defined as the solutions of the equations

$$u_n=R_k(RTu_n,\ QTu_{n-1})\qquad(u_0\in E,\ n=1,\ 2,\ \ldots), \tag{17.1}$$

where the operators R_k are defined by

$$R_1(PTx,\ QTy)=PTx+QTy,$$
$$R_i(PTx,\ QTy)=PTx+QTR_{i-1}(PTx,\ QTy)\quad(i=2,\ 3,\ \ldots,\ k). \tag{17.2}$$

In case (b), the equations for the successive approximations v_n are

$$v_n = S_k(Pv_n, Qv_{n-1}) \qquad (v_0 \in E, \quad n = 1, 2, \ldots), \tag{17.3}$$

with S_k defined by

$$\begin{aligned} &S_1(Px, Qy) = T(Px + Qy), \\ &S_i(Px, Qy) = T[Px + QS_{i-1}(Px, Qy)] \quad (i = 2, 3, \ldots, k). \end{aligned} \tag{17.4}$$

Finally, in case (c) we have successive approximations w_n defined by

$$w_n = U_k(PTw_n, Pw_n, Qw_{n-1}) \qquad (w_0 \in E, \; n = 1, 2, \ldots), \tag{17.5}$$

where the operators U_k are defined by

$$\begin{aligned} &U_1(PTx, Px, Qy) = PTx + QT(Px + Qy), \\ &U_i(PTx, Px, Qy) = PTx + QT[Px + QU_{i-1}(PTx, Px, Qy)] \\ &\qquad\qquad (i = 2, 3, \ldots, k). \end{aligned} \tag{17.6}$$

As shown in §9, in each of these three cases the procedure reduces to solution of operator equations in E_P and evaluation of certain operators.

Since we are dealing with orthogonal projections in Hilbert space, we can use (16.1) to establish convergence criteria and error estimates for these algorithms, which do not follow from the general theory of Chapter I.

In cases (a) and (b), the corresponding criteria and estimates are furnished by the following theorem.

THEOREM 17.1. *If the operators T, PT and QT satisfy Lipschitz conditions with constants q_T, q_{PT} and q_{QT} such that*

$$\min\left\{q_T^2 \sum_{i=0}^{k-1} q_{QT}^{2i}, \quad q_{PT}^2 + q_{QT}^2\right\} < 1, \tag{17.7}$$

then the successive approximations u_n and v_n are uniquely determined by equations (17.1) *and* (17.3) *for any $u_0 \in E$ and $v_0 \in E$, and converge to the unique solution of the equation $x = Tx$ in E; moreover, the following error estimates hold*:

$$\|x - u_n\| \leqslant N_k(1 - \varepsilon_k)^{-1}\varepsilon_k^{n-p}\|QTu_p - QTu_{p-1}\| \qquad (1 \leqslant p \leqslant n), \tag{17.8}$$

$$\|x - v_n\| \leqslant q_T N_k(1 - \varepsilon_k)^{-1}\varepsilon_k^{n-p}\|Qv_p - Qv_{p-1}\| \qquad (1 \leqslant p \leqslant n), \tag{17.9}$$

where

$$N_k = q_{QT}^{k-1}\left(1 - q_{PT}^2 \sum_{i=0}^{k-1} q_{QT}^{2i}\right)^{-\frac{1}{2}},$$

$$\varepsilon_k = \min\left\{q_T q_{QT}^{k-1},\ q_{QT}^k\left(1 - q_{PT}^2 \sum_{i=0}^{k-1} q_{QT}^{2i}\right)^{-\frac{1}{2}}\right\} < 1.$$

Moreover, $\epsilon_m < \epsilon_k$ *for* $m > k$, *if* $q_{QT} \neq 0$.

PROOF. That equations (17.1) and (17.3) are uniquely solvable for u_n and v_n follows from the fact that the operators in the right-hand sides of equation (17.1), and of the equation

$$Pv_n = PS_k(Pv_n, Qv_{n-1}), \tag{17.10}$$

obtained from (17.3) by applying the operator P to both sides, satisfy Lipschitz conditions with Lipschitz constants bounded above by the number

$$\Delta_k = q_{PT}\sqrt{\sum_{i=0}^{k-1} q_{QT}^{2i}},$$

which is less than 1 because of (17.7).

Indeed, setting $\delta z_n = z_n - z_{n-1}$ and using (16.1) (which is true because P is orthogonal), we deduce from (17.1) and (17.10) that

$$\| R_k(PTx, v) - R_k(PTy, v)\|^2 \leqslant q_{PT}^2 \sum_{i=0}^{k-1} q_{QT}^{2i} \| x - y\|^2,$$

$$\| PS_k(Px, v) - PS_k(Py, v)\|^2 \leqslant q_{PT}^2 \sum_{i=0}^{k-1} q_{QT}^{2i} \| Px - Py\|^2$$

for any $v \in E$. The inequality $\Delta_k < 1$ is easily derived from (17.7) if one notes that $q_{PT} \leqslant q_T$, and from the inequality $q_{PT}^2 + q_{QT}^2 < 1$.

We have

$$\Delta_k^2 \leqslant q_{PT}^2 \sum_{i=0}^{\infty} q_{QT}^{2i} = \frac{q_{PT}^2}{1 - q_{QT}^2} < 1.$$

We shall now show that the sequences $\{u_n\}$ and $\{v_n\}$ converge to solutions of the equations

$$x = R_k(PTx, QTx), \tag{17.11}$$

$$x = S_k(Px, Qx), \tag{17.12}$$

respectively, Setting $\delta Tu_n = Tu_n - Tu_{n-1}$ and using (16.1) we see from equations (17.1) and (17.3) that

$$\| Q\delta Tu_n\|^2 \leqslant q_{QT}^{2k} \| Q\delta Tu_{n-1}\|^2 + q_{QT}^2 \sum_{i=0}^{k-1} q_{QT}^{2i} \| P\delta Tu_n\|^2,$$

$$\| P\delta Tu_n\|^2 \leqslant q_{PT}^2 q_{QT}^{2(k-1)} \| Q\delta Tu_{n-1}\|^2 + q_{PT}^2 \sum_{i=0}^{k-1} q_{QT}^{2i} \| P\delta Tu_n\|^2$$

and

$$\| Q\delta v_n\|^2 \leqslant q_{QT}^{2k} \| Q\delta v_{n-1}\|^2 + q_{QT}^2 \sum_{i=0}^{k-1} q_{QT}^{2i} \| P\delta v_n\|^2,$$

$$\| P\delta v_n\|^2 \leqslant q_{PT}^2 q_{QT}^{2(k-1)} \| Q\delta v_{n-1}\|^2 + q_{PT}^2 \sum_{i=0}^{k-1} q_{QT}^{2i} \| P\delta v_n\|^2,$$

whence, since $\Delta_k^2 = q_{PT}^2 \sum_{i=0}^{k-1} q_{QT}^{2i} < 1$, we obtain

$$\| Q\delta T u_n \| \leqslant q_{QT}^k \left(1 - \Delta_k^2\right)^{-\frac{1}{2}} \| Q\delta T u_{n-1} \|, \tag{17.13}$$

$$\| Q\delta v_n \| \leqslant q_{QT}^k \left(1 - \Delta_k^2\right)^{-\frac{1}{2}} \| Q\delta v_{n-1} \|. \tag{17.14}$$

It also follows from (17.1) and (17.3) that

$$\| \delta T u_n \|^2 \leqslant q_T^2 q_{QT}^{2(k-1)} \| Q\delta T u_{n-1} \|^2 + q_T^2 \sum_{i=0}^{k-1} q_{QT}^{2i} \| P\delta T u_n \|^2, \tag{17.15}$$

$$\| \delta v_n \|^2 \leqslant q_T^2 q_{QT}^{2(k-1)} \| Q\delta v_{n-1} \|^2 + q_T^2 \sum_{i=0}^{k-1} q_{QT}^{2i} \| P\delta v_n \|^2. \tag{17.16}$$

But by (16.1)

$$\| \delta T u_n \|^2 = \| Q\delta T u_n \|^2 + \| P\delta T u_n \|^2, \qquad \| \delta v_n \|^2 = \| Q\delta v_n \|^2 + \| P\delta v_n \|^2.$$

Substituting these expressions on the left of (17.15) and (17.16), respectively, and then transferring $\|P\delta u_n\|^2$ and $\|P\delta v_n\|^2$ to the right and dropping the terms

$$\left(-1 + q_T^2 \sum_{i=0}^{k-1} q_{QT}^{2i}\right) \| P\delta T u_n \|^2, \qquad \left(-1 + q_T^2 \sum_{i=0}^{k-1} q_{QT}^{2i}\right) \| P\delta v_n \|^2,$$

if they are negative, we obtain

$$\| Q\delta T u_n \|^2 \leqslant q_T^2 q_{QT}^{2(k-1)} \| Q\delta T u_{n-1} \|^2, \tag{17.17}$$

$$\| Q\delta v_n \|^2 \leqslant q_T^2 q_{QT}^{2(k-1)} \| Q\delta v_{n-1} \|^2. \tag{17.18}$$

Comparing (17.17) with (17.13) and (17.18) with (17.14), we find that

$$\| Q\delta T u_n \| \leqslant \varepsilon_k \| Q\delta T u_{n-1} \|, \tag{17.19}$$

$$\| Q\delta v_n \| \leqslant \varepsilon_k \| Q\delta v_{n-1} \|, \tag{17.20}$$

whence it follows that $\{QTu_n\}$ and $\{Qv_n\}$ are Cauchy sequences. It is readily seen that $\{u_n\}$ and $\{v_n\}$ are also Cauchy sequences, and they converge to limits u and v which are solutions of (17.11) and (17.12) respectively.

But under our assumptions each of the equations $x = Tx$, (17.11) and (17.12) has a unique solution, and so $u = v = x$, i.e., u and v both coincide with the solution of the equation $x = Tx$.

To establish the error estimates (17.8) and (17.9), we need only observe that under the assumptions of the theorem

$$\| \delta u_n \|^2 \leqslant q_{QT}^{2(k-1)} \| Q\delta T u_{n-1} \|^2 + \sum_{i=0}^{k-1} q_{QT}^{2i} \| P\delta T u_n \|^2 \leqslant N_k^2 \| Q\delta T u_{n-1} \|^2,$$

$$\| \delta v_n \|^2 \leqslant q_T^2 q_{QT}^{2(k-1)} \| Q\delta v_{n-1} \|^2 + q_T^2 \sum_{i=0}^{k-1} q_{QT}^{2i} \| P\delta v_n \|^2 \leqslant q_T^2 N_k^2 \| Q\delta v_{n-1} \|^2.$$

We now consider the algorithms (17.1) and (17.3), replacing the operators P and Q by P' and Q' respectively, where P' is an orthogonal projection such that $PP' = P$, and $Q' = I - P'$. Then it is obvious that $PQ' = 0$, $P'Q = P' - P$ and

$QQ' = Q'$. For example, the operators P and P' defined by

$$Px = \sum_{i=1}^{l} (x, \varphi_i)\varphi_i, \qquad P'x = \sum_{i=1}^{m} (x, \varphi_i)\varphi_i, \quad m \geqslant l,$$

where $\{\varphi_i\}$ is an orthogonal system of elements of the Hilbert space E, satisfy these conditions.

We are thus considering the algorithms

$$u_n = R_k(P'Tu_n, Q'Tu_{n-1}) \qquad (u_0 \in E, \quad n = 1, 2, \ldots), \tag{17.21}$$

$$v_n = S_k(P'v_n, Q'v_{n-1}) \qquad (v_0 \in E, \quad n = 1, 2, \ldots), \tag{17.22}$$

where R_k and S_k are defined by (17.2) and (17.4) with P and Q replaced by P' and Q'.

We claim that if (17.21) and (17.22) are solvable for u_n and v_n and moreover

$$q_{PT}^2 + 2q_{QT}^2 < 1, \tag{17.23}$$

then algorithms (17.21) and (17.22) yield sequences $\{u_n\}$ and $\{v_n\}$ which converge to a solution of the equation $x = Tx$, and the rate of convergence is that of a geometric progression with quotient

$$q = \frac{q_{QT}^k}{\sqrt{1 - q_{PT}^2 \sum_{i=0}^{k-1} q_{QT}^{2i}}} < 1.$$

It follows from (17.21) and (17.22) that

$$\| Q\delta Tu_n \|^2 \leqslant q_{QT}^2 \sum_{i=1}^{k-1} q_{Q'T}^{2i} \| P'\delta Tu_n \|^2 + q_{QT}^2 q_{Q'T}^{2(k-1)} \| Q'\delta Tu_{n-1} \|,^2 \tag{17.24}$$

$$\| Q\delta v_n \|^2 \leqslant q_{QT}^2 \sum_{i=0}^{k-1} q_{Q'T}^{2i} \| P'\delta v_n \|^2 + q_{QT}^2 q_{Q'T}^{2(k-1)} \| Q'\delta v_{n-1} \|^2, \tag{17.25}$$

and also

$$\| P\delta Tu_n \|^2 \leqslant q_{PT}^2 \sum_{i=0}^{k-1} q_{Q'T}^{2i} \| P'\delta Tu_n \|^2 + q_{PT}^2 q_{Q'T}^{2(k-1)} \| Q'\delta Tu_{n-1} \|^2, \tag{17.26}$$

$$\| P\delta v_n \|^2 \leqslant q_{PT}^2 \sum_{i=0}^{k-1} q_{Q'T}^{2i} \| P'\delta v_n \|^2 + q_{PT}^2 q_{Q'T}^{2(k-1)} \| Q'\delta v_{n-1} \|^2. \tag{17.27}$$

Now, since $q_{Q'T} \leqslant q_{QT}$, $P' = P' - P + P$ and $(P' - P)P = 0$, it follows from (17.26) and (17.27) that

$$\| P\delta Tu_n \|^2 \leqslant q_{PT}^2 \sum_{i=0}^{k-1} q_{QT}^{2i} (\| (P' - P)\delta Tu_n \|^2 + \| P\delta Tu_n \|^2) + q_{PT}^2 q_{QT}^{2(k-1)} \| Q\delta Tu_{n-1} \|^2,$$

$$\| P\delta v_n \|^2 \leqslant q_{PT}^2 \sum_{i=0}^{k-1} q_{QT}^{2i} (\| (P' - P)\delta v_n \|^2 + \| P\delta u_n \|^2) + q_{PT}^2 q_{QT}^{2(k-1)} \| Q\delta v_{n-1} \|^2,$$

whence

$$\|P\delta Tu_n\|^2 \leqslant \left(1 - q_{PT}^2 \sum_{i=0}^{k-1} q_{QT}^{2i}\right)^{-1} q_{PT}^2 \left[\sum_{i=0}^{k-1} q_{QT}^{2i} \|(P' - P)\,\delta Tu_n\|^2 + q_{QT}^{2(k-1)} \|Q\delta Tu_{n-1}\|^2\right],$$

$$\|P\delta v_n\|^2 \leqslant \left(1 - q_{PT}^2 \sum_{i=0}^{k-1} q_{QT}^{2i}\right)^{-1} q_{PT}^2 \left[\sum_{i=0}^{k-1} q_{QT}^{2i} \|(P' - P)\,\delta v_n\|^2 + q_{QT}^{2(k-1)} \|Q\delta v_{n-1}\|^2\right].$$

Substituting these estimates for $\|P\delta Tu_n\|^2$ and $\|P\delta v_n\|^2$ into the inequalities

$$\|Q'\delta Tu_n\|^2 + \|(P' - P)\,\delta Tu_n\|^2 = \|QT\delta Tu_n\|^2$$
$$\leqslant q_{QT}^2 \sum_{i=0}^{k-1} q_{QT}^{2i} \left[\|(P' - P)\,\delta Tu_n\|^2 + \|P\delta Tu_n\|^2\right] + q_{QT}^{2k} \|Q'\delta Tu_{n-1}\|^2,$$

$$\|Q'\delta v_n\|^2 + \|(P' - P)\,\delta v_n\|^2 = \|Q\delta v_n\|^2$$
$$\leqslant q_{QT}^2 \sum_{i=0}^{k-1} q_{QT}^{2i} \left[\|(P' - P)\,\delta v_n\|^2 + \|P\delta v_n\|^2\right] + q_{QT}^{2k} \|Q'\delta v_{n-1}\|^2,$$

which are easily derived from (17.24) and (17.25) if we note that $Q = Q' + P' - P$ and $Q'(P' - P) = 0$, while $q_{Q'T} \leqslant q_{QT}$, we see that

$$\|Q'\delta Tu_n\|^2 + \|(P' - P)\,\delta Tu_n\|^2 \leqslant \frac{\sum_{i=1}^{k} q_{QT}^{2i}}{1 - q_{PT}^2 \sum_{i=0}^{k-1} q_{QT}^{2i}} \|(P' - P)\,\delta Tu_n\|^2 + q^2 \|Q'\delta Tu_{n-1}\|^2, \tag{17.28}$$

$$\|Q'\delta v_n\|^2 + \|(P' - P)\,\delta v_n\|^2 \leqslant \frac{\sum_{i=1}^{k} q_{QT}^{2i}}{1 - q_{PT}^2 \sum_{i=0}^{k-1} q_{QT}^{2i}} \|(P' - P)\,\delta v_n\|^2 + q^2 \|Q'\delta v_{n-1}\|^2. \tag{17.29}$$

Since by condition (17.23)

$$\sum_{i=1}^{k} q_{QT}^{2i} \left(1 - q_{PT}^2 \sum_{j=0}^{k-1} q_{QT}^{2j}\right)^{-1} < 1$$

for any $k = 1, 2, \ldots$, it follows that (17.28) and (17.29) may be improved, replacing them by

$$\|Q'\delta Tu_n\| \leqslant q^2 \|Q'\delta Tu_{n-1}\|^2, \qquad \|Q'\delta v_n\|^2 \leqslant q^2 \|Q'\delta v_{n-1}\|^2,$$

and it follows from these inequalities that $\{Q'Tu_n\}$ and $\{Q'v_n\}$ are Cauchy sequences. The rest of the proof that the sequences $\{u_n\}$ and $\{v_n\}$ converge to a solution of the equation $x = Tx$ is analogous to the reasoning in the proof of Theorem 17.1.

We now consider algorithm (17.5).

THEOREM 17.2. *Suppose that the operators PT and QT satisfy Lipschitz conditions with constants* q_{PT} *and* q_{QT} *such that*

$$q_{PT}^2 + q_{QT}^2 < 1. \tag{17.30}$$

Then equations (17.5) *are uniquely solvable for* $w_n \in E$ *and the sequence* $\{w_n\}$ *converges to a solution of the equation* $x = Tx$, *with error estimate*

$$\| x - w_n \| \leqslant \frac{1}{\sqrt{1 - q_{PT}^2}} \cdot \frac{1}{1 - \varepsilon_k} \varepsilon_k^{n-p+1} \| Q (w_p - w_{p-1}) \| \tag{17.31}$$

$$(1 \leqslant p \leqslant n),$$

where

$$\varepsilon_k^2 = \frac{q_{QT}^{2k}}{1 - \overline{\Delta}_k^2} < 1, \quad \overline{\Delta}_k^2 = \frac{q_{PT}^2 \sum_{i=1}^{k} q_{QT}^{2i}}{1 - q_{PT}^2} < 1.$$

Moreover, $\epsilon_m < \epsilon_k$ *for* $m > k$, *if* $q_{QT} \neq 0$.

PROOF. That equations (17.5) are uniquely solvable for w_n follows from the fact that the operator on the right of the equations satisfies a Lipschitz condition with constant less than 1. Since $Pw_n = PTw_n$, we have

$$\| U_k (PTx, PTx, Qz) - U_k (PTy, PTy, Qz) \|^2 \leqslant q_{PT}^2 \sum_{i=0}^{k} q_{QT}^{2i} \| x - y \|^2,$$

where, by (17.30),

$$q_{PT}^2 \sum_{i=0}^{k} q_{QT}^{2i} < \frac{q_{PT}^2}{1 - q_{QT}^2} < 1.$$

Setting $\delta_n = w_n - w_{n-1}$, we deduce from (17.5) that

$$\| Q\delta_n \|^2 \leqslant \sum_{i=1}^{k} q_{QT}^{2i} \| P\delta_n \|^2 + q_{QT}^{2k} \| Q\delta_{n-1} \|^2. \tag{17.32}$$

It also follows from (17.5) that

$$\| P\delta_n \|^2 = \| PTw_n - PTw_{n-1} \|^2 \leqslant q_{PT}^2 \| \delta_n \|^2 = q_{PT}^2 (\| P\delta_n \|^2 + \| Q\delta_n \|^2),$$

whence

$$\| P\delta_n \|^2 \leqslant \frac{q_{PT}^2}{1 - q_{PT}^2} \| Q\delta_n \|^2.$$

Substituting this in the right-hand side of (17.32), we obtain

$$\| Q\delta_n \|^2 \leqslant \overline{\Delta}_k^2 \| Q\delta_n \|^2 + q_{QT}^{2k} \| Q\delta_{n-1} \|^2.$$

It is readily seen that because of condition (17.30) $\overline{\Delta}_k^2 < 1$, and so $\|Q\delta_n\|^2 \leqslant \epsilon_k^2 \|Q\delta_{n-1}\|^2$. Since $\epsilon_k^2 < 1$ (again by (17.30)), $\{Qw_n\}$ is a Cauchy sequence. Since $Pw_n = PTw_n$, we have

$$\delta_n = P\delta_n + Q\delta_n = PTw_n - PTw_{n-1} + Q\delta_n$$

and $\|\delta_n\|^2 \leqslant q_{PT}^2 \|\delta_n\|^2 + \|Q\delta_n\|^2$, so that

$$\| \delta_n \|^2 \leqslant \frac{1}{1 - q_{PT}^2} \| Q\delta_n \|^2.$$

Thus $\{w_n\}$ is also a Cauchy sequence, and so it has a limit, which is a solution of the equation $x = Tx$.

The error estimate (17.31) is proved in the usual manner.

We now consider the case that the assumptions of Theorems 17.1 and 17.2 hold not in the whole space E but only in some subset.

As remarked in §9, if it is true that for any x and y in some closed subset D of E

$$PTx + QTy \in D, \tag{17.33}$$

then $R_i(x, y) \in D$ for any $i = 1, 2, \ldots$. Consequently, if the assumptions of Theorem 17.1 are satisfied for $x, y \in D$, then equations (17.1) are uniquely solvable for u_n in D, and the sequence of solutions $\{u_n\}$ converges to the unique solution in D of the equation $x = Tx$, in such a way that the error estimate (17.8) holds.

We give some examples of sets for which condition (17.33) holds.

I. Let D be the ball $S(g, r)$. If

$$\| Tg - g \| + \sqrt{q_{PT}^2 + q_{QT}^2}\, r \leqslant r, \tag{17.34}$$

then $PTx + QTy \in S(g, r)$ for any $x, y \in S(g, r)$. Indeed, for any $x, y \in S(g, r)$

$$\begin{aligned} \| PTx + QTy - g \| &= \| PTx - PTg + QTy - QTg + PTg + QTg - g \| \\ &\leqslant \sqrt{\| PTx - PTg \|^2 + \| QTy - QTg \|^2} + \| Tg - g \| \\ &\leqslant \sqrt{q_{PT}^2 + q_{QT}^2}\, r + \| Tg - g \| \leqslant r, \end{aligned}$$

so that $PTx + QTy \in S(g, r)$.

II. Suppose that for all $x, y \in S(g, r)$, with r as yet undetermined,

$$\| PTx - Pg \| \leqslant V_{PT}(r), \quad \| QTy - Qg \| \leqslant V_{QT}(r), \tag{17.35}$$

where $V_{PT}(r)$ and $V_{QT}(r)$ are certain functions. If there is a number r^* such that

$$V_{PT}^2(r) + V_{QT}^2(r) \leqslant r^2, \tag{17.36}$$

then clearly $PTx + QTy \in S(g, r^*)$ for $x, y \in S(g, r^*)$.

III. Suppose that for all $x, y \in S(g, r)$, where r is to be determined,

$$\begin{aligned} \| PTx + QTy - PTg - QTg \|^2 \leqslant W_{PT}^2(r) \| x - g \|^2 & \\ + W_{QT}^2(r) \| y - g \|^2, & \end{aligned} \tag{17.37}$$

where $W_{PT}(r)$ and $W_{QT}(r)$ are certain functions. If there exists a positive r^* such that

$$\|Tg-g\|+\sqrt{W_{PT}^2(r)+W_{QT}^2(r)}\,r\leqslant r, \tag{17.38}$$

then $PTx + QTy \in S(g, r^*)$ for $x, y \in S(g, r^*)$.

The proof is analogous to the proof in case I.

We consider a few more variants of algorithms (17.3) and (17.5).

As remarked in §9, if it is true that for all x, y in a subset D of E

$$PTx+Qv_0\in D, \tag{17.39}$$

$$PTx+QTy\in D, \tag{17.40}$$

then

$$PTx+QTR_{i-1}(PTx+Qv_{n-1})\in D. \tag{17.41}$$

Using the fact that $v_n = Tz_n$, where $z_n = Pv_n + QS_{k-1}(Pv_n, Qv_{n-1})$, one readily proves the following statement: if D is a closed set and for all $x, y \in D$ the assumptions of Theorem 17.1 are satisfied, then the equations

$$\begin{aligned} z_1 &= R_k(PTz, Qv_0),\\ z_n &= R_k(PTz_n, QTz_{n-1}) \qquad (n=2,\ 3,\ \ldots) \end{aligned} \tag{17.42}$$

are uniquely solvable in D for z_n $(n = 1, 2, \ldots)$ and the sequences $\{z_n\}$ and $\{v_n\}$, where z_n are the solutions of (17.42), converge to a solution of the equation $x = Tx$ satisfying the error estimate (17.9).

The proof is similar to that of Theorem 17.1, except that instead of E one restricts attention to the set D.

A similar situation occurs for the algorithm (17.5). If for all $x, y \in D$

$$PTx+Qw_0\in D \tag{17.43}$$

and (17.40) holds, then

$$U_i(PTx,\ Px,\ Qw_{n-1})\in D \tag{17.44}$$

for all $n = 1, 2, \ldots$. Therefore, if D is closed and the assumptions of Theorem 17.2 hold for all $x, y \in D$, then equations (17.5) are uniquely solvable for w_n in D and the sequence of solutions $\{w_n\}$ converges to the unique solution in D of the equation $x = Tx$, with error estimate (17.31).

If D is a ball $S(g, r)$, then conditions (17.39), (17.40), and also (17.43), (17.40) with v_0 replaced by w_0, are satisfied, e.g., in the cases I–III listed above except that inequalities (17.34), (17.36) and (17.38) should be replaced by the following systems of inequalities:

$$\|Qv_0+PTg-g\|+q_{PT}r\leqslant r, \qquad \|Tg-g\|+\sqrt{q_{PT}^2+q_{QT}^2}\,r\leqslant r; \tag{17.45}$$

$$\|Q(v_0-g)\|^2+V_{PT}^2(r)\leqslant r^2, \qquad V_{PT}^2(r)+V_{QT}^2(r)\leqslant r^2; \tag{17.46}$$

$$\|Qv_0 + PTg - g\| + W_{PT}(r) \leqslant r,$$
$$\|Tg - g\| + \sqrt{W_{PT}^2(r) + W_{QT}^2(r)}\, r \leqslant r. \tag{17.47}$$

§18. Investigation of modified generalized iterative projection method for operator equations in Hilbert space

We again consider the operator equation $x = Tx$ for an operator T defined on a Hilbert space E.

Let P be an orthogonal projection of E onto a subspace E_P.

Consider the algorithm studied in §§11 and 12 for lattice-normed spaces and in §15 for Banach spaces, according to which the successive approximations x_n to the solution of the equation $x = Tx$ are defined by $x_n = u_n + v_n$, where u_n and v_n are solutions of the systems of operator equations

$$\begin{aligned} u_n &= PT(u_n + v_n), \\ v_n &= QS_k(u_n, v_{n-1}) \quad (n = 1, 2, \ldots); \end{aligned} \tag{18.1}$$

the operators S_k being defined recursively by

$$\begin{aligned} S_1(u, v) &= QT(u + v), \\ S_i(u, v) &= QT[u + S_{i-1}(u, v)], \end{aligned} \tag{18.2}$$

with v_0 an arbitrary element of the space $E - E_P = E_Q$.

We shall derive some convergence criteria for this algorithm which utilize the special properties of orthogonal projections.

THEOREM 18.1. *Suppose that the operators PT and QT satisfy Lipschitz conditions with constants q_{PT} and q_{QT} such that*

$$q_{PT}^2 + q_{QT}^2 < 1. \tag{18.3}$$

Then the system of operator equations (18.1) *is uniquely solvable for u_n and v_n, and the sequence $\{x_n\}$, $x_n = u_n + v_n$, converges to a solution of the equation $x = Tx$, with error estimate*

$$\|x - x_n\| \leqslant \frac{1}{\sqrt{1 - q_{PT}^2}} \cdot \frac{\varepsilon_k^{n-p+1}}{1 - \varepsilon_k} \|Q\delta v_p\| \tag{18.4}$$
$$(\delta v_p = v_p - v_{p-1},\ 1 \leqslant p \leqslant n),$$

where

$$\varepsilon_k = \frac{\sqrt{1 - q_{PT}^2}\, q_{QT}^k}{\sqrt{1 - q_{PT}^2 \sum_{i=0}^{k} q_{QT}^{2i}}} < 1. \tag{18.5}$$

Moreover, $\epsilon_m < \epsilon_k$ for $m > k$, if $q_{QT} \neq 0$.

PROOF. Substituting the expression for v_n from the second equation of (18.1) into the first, we obtain

$$Pu_n = PT\,[Pu_n + QS_k\,(Pu_n, Qv_{n-1})]. \tag{18.6}$$

Using the identity

$$\| Px + Qy \|^2 = \| Px \|^2 + \| Qy \|^2, \tag{18.7}$$

we get

$$\| PT\,[Px + QS_k\,(Px, Qz)] - PT\,[Py + QS_k\,(Py, Qz)] \|^2$$
$$\leqslant q_{PT}^2 \sum_{i=0}^{k} q_{QT}^{2i} \| x - y \|^2;$$

in other words, the operator on the right of (18.6) satisfies a Lipschitz condition with constant

$$\Delta_k = q_{PT} \sqrt{\sum_{i=0}^{k} q_{QT}^{2i}}.$$

But, by (18.3), $\Delta_k < 1$ and so we can apply the contraction mapping principle to equation (18.6). Consequently, equations (18.6) have unique solutions Pu_n for any $n = 1, 2, \ldots$. Substituting these solutions on the right of the second equation of (18.1), we find v_n. Thus the sequence $\{x_n\}$ is uniquely determined.

We now show that $\{x_n\}$ is a Cauchy sequence. It follows from (18.6) that

$$\| P\delta u_n \|^2 \leqslant \Delta_k^2 \| P\delta u_n \|^2 + q_{PT}^2 q_{QT}^{2k} \| Q\delta v_{n-1} \|^2,$$

whence

$$\| P\delta u_n \|^2 \leqslant \frac{q_{PT}^2 q_{QT}^{2k}}{1 - \Delta_k^2} \| Q\delta v_{n-1} \|^2. \tag{18.8}$$

By the second equation of (18.1),

$$\| Q\delta v_n \|^2 \leqslant \sum_{i=1}^{k} q_{QT}^{2i} \| P\delta u_n \|^2 + q_{QT}^{2k} \| Q\delta v_{n-1} \|^2.$$

Substituting the estimate for $\|P\delta u_n\|$ on the right of this inequality, we get

$$\| Q\delta v_n \|^2 \leqslant \varepsilon_k^2 \| Q\delta v_{n-1} \|^2, \tag{18.9}$$

whence it follows that $\{Qv_n\}$ is a Cauchy sequence, since it is readily deduced from (18.3) that $\epsilon_k < 1$. In view of (18.8), we see that $\{Pu_n\}$ is also a Cauchy sequence, and therefore both $\{v_n\}$ and $\{u_n\}$ converge to limits v and u, respectively, which satisfy the system of equations

$$u = PT\,(u + v), \qquad v = QS_k\,(u, v). \tag{18.10}$$

As in §11, one can now prove that u, v is the unique solution of this system. And since for $k = 1$ system (18.10) has a unique solution (assuming (18.3)) and any solution of system (18.10) ($k = 1$) is also a solution of this system for $k > 1$, it follows that u, v is a solution of system (18.10) for $k = 1$. Hence it follows that $x = u + v$ is a solution of the equation $x = Tx$.

In order to establish the error estimate (18.4), we note that $x_n = R_{PT} v_n$

(where R_{PT} is the resolvent operator for PT). We obtain

$$\|\delta x_n\|^2 \leqslant q_{PT}^2 \|\delta x_n\|^2 + \|Q\delta v_n\|^2 \quad \text{and} \quad \|\delta x_n\| \leqslant \frac{1}{\sqrt{1-q_{PT}^2}} \|Q\delta v_n\|.$$

We now assume that the conditions of Theorem 18.1 hold not throughout the space but only in some subset of E.

It follows at once from the results of §11 that if D_1 and D_2 are arbitrary closed subsets of E and D is the set of elements $x = u + v$, where $u \in D_1, v \in D_2$, and for all $x, y \in D$,

$$PTx \in D_1, \ QTy \in D_2, \tag{18.11}$$

then $PT(u + v) \in D_1$ and $S_i(u, v) \in D_2$ for $u \in D_1$ and $v \in D_2$.

Hence it follows in particular that, if (18.11) is satisfied for any $x, y \in D$ and the assumptions of Theorem 18.1 are valid for all elements of D, then for all n the system (18.1) has a unique solution u_n, v_n, with $u_n \in D_1, v_n \in D_2$, provided $v_0 \in D_2$; moreover, the successive approximations $x_n = u_n + v_n$ converge to the unique solution in D of the equation $x = Tx$ with error estimate (18.4).

We consider a few special cases in which (18.11) is true for $x, y \in D$.

I. Suppose that for all $x, y \in S(g, r)$, where g is a fixed element and r is to be determined,

$$\begin{aligned} \|PTx - Pg\| &\leqslant V_{PT}(r), \\ \|QTy - Qg\| &\leqslant V_{QT}(r). \end{aligned} \tag{18.12}$$

If there exists a positive r^* such that

$$V_{PT}^2(r) + V_{QT}^2(r) \leqslant r^2, \tag{18.13}$$

then for any $x, y \in S(g, r^*)$

$$PTx \in S(Pg, V_{PT}(r^*)), \ QTy \in S(Qg, V_{QT}(r^*))$$

and the sets D, D_1 and D_2 will be, respectively,

$$S(g, r^*), \ S(Pg, V_{PT}(r^*)) \ \text{and} \ S(Qg, V_{QT}(r^*)).$$

To prove this, we need only observe that if $u \in D_1$ and $v \in D_2$, then $x = u + v \in D$, since

$$\|Pu + Qv - g\|^2 = \|Pu - Pg\|^2 + \|Qv - Qg\|^2 \leqslant V_{PT}^2(r^*) + V_{QT}^2(r^*) \leqslant r^*.$$

II. Suppose that for all $x, y \in S(g, r)$, where g is a known element of E and r is to be determined,

$$\|PTx - PTg\| \leqslant W_{PT}(r), \qquad \|QTy - QTg\| \leqslant W_{QT}(r), \tag{18.14}$$

where $W_{PT}(r)$ and $W_{QT}(r)$ are certain functions.

If there exists $r^* > 0$ such that

$$[\|PTg - Pg\| + W_{PT}(r)\, r]^2 + [\|QTg - Qg\| + W_{QT}(r)\, r]^2 \leqslant r^2, \tag{18.15}$$

then for all $x, y \in S(g, r^*)$

$$PTx \in S(Pg, r_1),\ QTy \in S(Qg, r_2), \tag{18.16}$$

where $r_1 = \|PTg - Pg\| + W_{PT}(r^*)\, r^*$, $\quad r_2 = \|QTg - Qg\| + W_{QT}(r^*)\, r^*$

and $u + v \in S(g, r^*)$ for $u \in S(Pg, r_1)$, $v \in S(Qg, r_2)$, $u \in E_P$ and $v \in E_Q$.

The proof of (18.16) follows the same lines as at the end of §11. We need only set $U = PT$, $S = QT$, $z_1 = Pg$, $z_2 = Qg$ and $z = g$.

Further, if $u \in S(Pg, r_1)$ and $v \in S(Qg, r_2)$, then

$$\|u + v - g\|^2 = \|u - Pg + v - Qg\|^2 = \|Pu - Pg\|^2 + \|Qv - Qg\|^2$$
$$\leqslant [\|PTg - Pg\| + W_{PT}(r^*)\, r^*]^2 + [\|QTg - Qg\| + W_{QT}(r^*)\, r^*]^2 \leqslant r^*,$$

so that $u + v \in S(g, r^*)$.

§19. Iterative projection method based on minimization of the norm of the residual

The iterative projection method studied in this section, for solution of operator equations (linear or nonlinear) in a Hilbert space, is a modification and generalization of the method of averaged functional corrections. We shall propose a general scheme which includes the algorithms of [129], [130], [132] and [133] as special cases, and establish convergence criteria and error estimates.

1. Consider the operator equation

$$x = Tx, \tag{19.1}$$

where T is an operator in a Hilbert space H and x is an unknown element of H.

Let $\{\varphi_{i,n}\}$ $(i = 1, \ldots, k(n))$ and $\{\psi_{i,n}\}$ $(i = 1, \ldots, l(n))$, where n is a fixed natural number, be two systems of linearly independent elements of H. The idea of the iterative projection method to be considered is to determine successive approximations x_n to the solution of (19.1) through the formula

$$x_n = T\left[x_{n-1} + \sum_{i=1}^{k(n)} C_{i,n}\varphi_{i,n}\right] + \sum_{i=1}^{l(n)} D_{i,n}\psi_{i,n} \quad (n = 1, 2, \ldots), \tag{19.2}$$

choosing the real constants $C_{i,n}$ and $D_{i,n}$ so that the norm of the residual $\epsilon(x_n) = x_n - Tx_n$ (where x_0 is an arbitrary element of H) is minimal. The residuals $\epsilon(x_n)$ are given by

$$\begin{aligned} \varepsilon(x_n) &= \varepsilon_n[C_{1,n}, \ldots, C_{k(n),n};\ D_{1,n}, \ldots, D_{l(n),n}] \\ &= T\left[x_{n-1} + \sum_{i=1}^{k(n)} C_{i,n}\varphi_{i,n}\right] + \sum_{i=1}^{l(n)} D_{i,n}\psi_{i,n} \\ &\quad - T\left\{T\left[x_{n-1} + \sum_{i=1}^{k(n)} C_{i,n}\varphi_{i,n}\right] + \sum_{i=1}^{l(n)} D_{i,n}\psi_{i,n}\right\}. \end{aligned} \tag{19.3}$$

We shall assume that $\|\epsilon(x_n)\|$ is a differentiable function of $C_{i,n}$ and $D_{j,n}$ ($i = 1, \ldots, k(n)$; $j = 1, \ldots, l(n)$). Since the existence of an extremum implies the vanishing of the first partial derivatives with respect to all the independent variables, we can determine the constants $C_{i,n}$ and $D_{i,n}$ at each step from the following system of $K(n) = k(n) + l(n)$ algebraic or transcendental equations in $K(n)$ unknowns:

$$\begin{gathered}\frac{\partial \| \varepsilon_n [C_{1,n}, \ldots, C_{k(n),n}; D_{1,n}, \ldots, D_{l(n),n}] \|}{\partial C_{i,n}} = 0, \\ \frac{\partial \| \varepsilon_n [C_{1,n}, \ldots, C_{k(n),n}; D_{1,n}, \ldots, D_{l(n),n}] \|}{\partial D_{j,n}} = 0\end{gathered} \tag{19.4}$$

$$(i = 1, 2, \ldots, k(n); \; j = 1, 2, \ldots, l(n)).$$

THEOREM 19.1. *If the operator T satisfies a Lipschitz condition*

$$\| Tx - Ty \| \leqslant q_T \| x - y \| \quad (x, y \in H) \tag{19.5}$$

with constant $q_T < 1$, then the successive approximations defined as above converge to the unique solution x in H of equation (19.1), *with the error estimate*

$$\| x - x_n \| \leqslant (1 - q_T)^{-1} q_T^{n-p} \| \varepsilon (x_p) \| \quad (0 \leqslant p \leqslant n). \tag{19.6}$$

PROOF. Let $\overline{C}_{1,n}, \ldots, \overline{C}_{k(n),n}$ and $\overline{D}_{1,n}, \ldots, \overline{D}_{l(n),n}$ be values of the constants $C_{1,n}, \ldots, C_{k(n),n}$ and $D_{1,n}, \ldots, D_{l(n),n}$ that minimize the function $\|\epsilon(x_n)\|$. Then, by (19.3) and (19.5),

$$\begin{aligned}\| \varepsilon_n [\overline{C}_{1,n}, \ldots, \overline{C}_{k(n),n}; \overline{D}_{1,n}, \ldots, \overline{D}_{l(n),n}] \| &\leqslant \| \varepsilon_n [0, \ldots, 0; 0, \ldots, 0] \| \\ &= \| Tx_{n-1} - TTx_{n-1} \| \leqslant q_T \| \varepsilon (x_{n-1}) \|.\end{aligned} \tag{19.7}$$

Applying inequality (19.7) successively for $n = 1, 2, \ldots$, we get

$$\| \varepsilon (x_n) \| \leqslant q_T^{n-p} \| \varepsilon (x_p) \|, \tag{19.8}$$

and so, since $q_T < 1$, $\lim_{n \to \infty} \|\epsilon(x_n)\| = 0$, i.e., the sequence $\{x_n\}$ converges to a solution of (19.1).

The error estimate (19.6) follows from (19.8) and the obvious inequality

$$\| x - x_n \| \leqslant \| Tx - Tx_n \| + \| Tx_n - x_n \| \leqslant q_T \| x - x_n \| + \| \varepsilon (x_n) \|.$$

COROLLARY. *Suppose that for each n the constants $\overline{C}_{1,n}, \ldots, \overline{C}_{k(n),n}$ and $\overline{D}_{1,n}, \ldots, \overline{D}_{l(n),n}$ minimizing $\|\epsilon(x_n)\|$ are such that x_n and Tx_n remain within some closed subset D of H, and that condition* (19.5) *holds in this subset. Then the successive approximations x_n converge to the unique solution $x \in D$ of equation* (19.1).

We now enlarge the number of elements in the systems $\{\varphi_{i,n}\}$ and $\{\psi_{i,n}\}$, i.e., add new elements at each step.

THEOREM 19.2. *Assume that for any n the systems* $\{\varphi_{i,n}\}(i = 1, \ldots, k(n))$ *and* $\{\psi_{i,n}\}$ $(i = 1, \ldots, l(n))$ *have the property that for arbitrary* $y \in H$

$$\|\varepsilon(x_n)\| \leqslant q \|\varepsilon(y)\| \quad (q < 1), \tag{19.9}$$

where $\epsilon(y) = y - Ty$ *and* $\epsilon(x_n) = x_n - Tx_n$, *where* x_n *is defined by* (19.2) *and* (19.4) *with* $x_{n-1} = y$.

Then the algorithm will also converge to the solution of equation (19.1) *when new elements are added to the systems* $\{\varphi_{i,n}\}$ *and* $\{\psi_{i,n}\}$. *Moreover, the residuals corresponding to the new successive approximations converge in norm at least as rapidly as a geometric progression with quotient* q.

PROOF. Let x_0 be the initial approximation, By (19.9), the first approximation x_1, based on the original set of elements, satisfies the condition

$$\|\varepsilon(x_1)\| \leqslant q \|\varepsilon(x_0)\|. \tag{19.10}$$

The first approximation $\bar{x}_1$ based on the enlarged systems will satisfy the inequality

$$\|\varepsilon(\bar{x}_1)\| \leqslant \|\varepsilon(x_1)\|, \tag{19.11}$$

since increasing the number of variables cannot increase the minimum of $\|\epsilon_n\|$. Comparing (19.10) and (19.11), we see that

$$\|\varepsilon(\bar{x}_1)\| \leqslant q \|\varepsilon(x_0)\|. \tag{19.12}$$

Next, taking $\bar{x}_1$ as the initial approximation, we find

$$\|\varepsilon(\bar{x}_2)\| \leqslant q \|\varepsilon(\bar{x}_1)\|, \tag{19.13}$$

where $\bar{x}_2$ corresponds to the enlarged set of elements and the initial approximation $\bar{x}_1$. Reasoning in analogous fashion, we see that in general

$$\|\varepsilon(\bar{x}_n)\| \leqslant q \|\varepsilon(\bar{x}_{n-1})\| \quad (n = 1, 2, \ldots), \tag{19.14}$$

and this proves the theorem.

2. Let us now assume that equation (19.1) is linear, i.e. $Tx = f + Ax$, where A is a linear operator in H and f some element of H. Equation (19.1) is now

$$x = f + Ax, \tag{19.15}$$

and we can write (19.2) as

$$x_n = f + Ax_{n-1} + \sum_{i=1}^{k(n)} C_{i,n} A\varphi_{i,n} + \sum_{i=1}^{l(n)} D_{i,n}\psi_{i,n}. \tag{19.16}$$

Suppose that 1 is not an eigenvalue of the operator A. To simplify the notation, we set

$$K(n) = k(n) + l(n),$$

$$\Phi_{i,n} = \begin{cases} A(I - A)\varphi_{i,n} & (i = 1, 2, \ldots, k(n)), \\ (I - A)\psi_{i-k(n),n} & (i = k(n) + 1, \ldots, k(n) + l(n)); \end{cases}$$

$$E_{i,n} = \begin{cases} C_{i,n} & (i = 1, 2, \ldots, k(n)), \\ D_{i-k(n),n} & (i = 1, 2, \ldots, k(n) + l(n)). \end{cases}$$

Then the expression for $\epsilon(x_n)$ is

$$\varepsilon(x_n) = A\varepsilon(x_{n-1}) + \sum_{i=1}^{K(n)} E_{i,n}\Phi_{i,n}, \tag{19.17}$$

and thus at each step determination of the constants $E_{i,n}$ minimizing $\|\epsilon(x_n)\|$ involves solution of $K(n)$ linear algebraic equations in $K(n)$ unknowns:

$$\sum_{i=1}^{K(n)} E_{i,n}(\Phi_{i,n}, \Phi_{j,n}) + (A\varepsilon(x_{n-1}), \Phi_{j,n}) = 0 \quad (j = 1, \ldots, K(n)) \tag{19.18}$$

(where (x, y) denotes the scalar product of $x, y \in H$).

If the system $\{\Phi_{i,n}\}$ $(i = 1, \ldots, K(n))$ is linearly independent, then the determinant of system (19.18), as the Gram determinant of a system of linearly independent elements, does not vanish; consequently (19.18) will have a unique solution.

The second term on the right of (19.17) is the projection of the element $-A\epsilon(x_{n-1})$ on the subspace spanned by the elements $\Phi_{i,n}$. Denote this projection operator by $P_{K(n)}$. We have

$$-P_{K(n)}A\varepsilon(x_{n-1}) = \sum_{i=1}^{K(n)} E_{i,n}\Phi_{i,n}, \tag{19.19}$$

where the coefficients $E_{i,n}$ are determined from (19.18), and we may write (19.17) as

$$\varepsilon(x_n) = Q_{K(n)}A\varepsilon(x_{n-1}), \; Q_{K(n)} = I - P_{K(n)}. \tag{19.20}$$

Note that the norm of each of the operators $P_{K(n)}$ and $Q_{K(n)}$ is unity, since $P_{K(n)}$ and $Q_{K(n)}$ are orthogonal projections.

We may now rewrite (19.20) as

$$\varepsilon(x_n) = Q_{K(n)}AQ_{K(n-1)}A \ldots Q_{K(1)}A\varepsilon(x_0). \tag{19.21}$$

Hence it follows that if

$$\lim_{n\to\infty} \| Q_{K(n)}AQ_{K(n-1)}A \ldots Q_{K(1)}A \| = 0, \tag{19.22}$$

then the algorithm converges to a solution of equation (19.15). Moreover, since $x - x_n = Ax - Ax_n - \epsilon(x_n)$, we then have

$$\| x - x_n \| \leqslant \| (I - A)^{-1} \| \| \varepsilon(x_n) \|. \tag{19.23}$$

Now suppose that the systems $\{\varphi_{i,n}\}$ and $\{\psi_{i,n}\}$ are independent of n, so that each step involves the same two systems, which we now denote by

$\{\varphi_i\}$ $(i = 1, \ldots, k)$ and $\{\psi_i\}$ $(i = 1, \ldots, l)$, respectively. Then the corresponding projections are also independent of n: $P_{K(n)} = P_K$ $(K = k + l)$. In this case we have the following two theorems.

THEOREM 19.3. *If*

$$\| Q_K A Q_K \| < 1 \quad (Q_K = I - P_K), \tag{19.24}$$

then the above successive approximation process converges to a solution of equation (19.15).

PROOF. This follows from (19.20) by application of the operator $Q_K A$, for then

$$Q_K A \varepsilon (x_n) = Q_K A Q_K Q_K A \varepsilon (x_{n-1}),$$

whence

$$\| Q_K A \varepsilon (x_n) \| \leqslant \| Q_K A Q_K \|^{n-p} \| Q_K A \varepsilon (x_p) \| (0 \leqslant p \leqslant n). \tag{19.25}$$

Consequently, if (19.24) holds the sequence $\|Q_K A \varepsilon(x_n)\|$ converges to zero, proving the theorem.

THEOREM 19.4. *If inequality* (19.24) *holds, the above successive approximation process also converges to a solution of equation* (19.15) *when new elements are added to the existing* φ_i $(i = 1, \ldots, k)$ *and* ψ_i $(i = 1, \ldots, l)$. *Then*

$$\| Q_{K+m} A Q_{K+m} \| \leqslant \| Q_K A Q_K \|, \tag{19.26}$$

where P_{K+m} *and* $Q_{K+m} = I - P_{K+m}$ *correspond to the new projection operator (for the enlarged system).*

PROOF. The proof is based on the equality $P_K P_{K+m} = P_{K+m} P_K = P_K$, which implies that

$$Q_{K+m} A Q_{K+m} = Q_{K+m} Q_K A Q_K Q_{K+m}.$$

But since $\|Q_{K+m}\| = 1$, this at once yields (19.26).

§20. Iterative projection method based on minimization of the norm of the error

This section is devoted to an iterative projection method designed specially for the treatment of linear operator equations in a Hilbert space H.

Consider the linear equation

$$x = f + Ax, \tag{20.1}$$

where A is a linear operator in a Hilbert space H and f is a known element of H.

We shall assume that 1 is a regular value of the operator A, so that equation (20.1) has a unique solution in H.

Let P be some projection operator mapping H onto a subspace H_P of finite or infinite dimension. As before, Q will denote the difference $I - P$, where I is the identity operator. Equation (20.1) may be written

$$x = f + PAx + QAx. \tag{20.2}$$

Since by (20.1) we have $x = (I - A)^{-1}f$, substitution of this value of x into the second term on the right of (20.2) gives

$$x = f + PA(I - A)^{-1}f + QAx. \tag{20.3}$$

We at once observe that if P is an arbitrary projection the element $PA(I - A)^{-1}f$ cannot be found, and so in practice we cannot always pass from (20.1) to (20.3). Nevertheless, by a special choice of P we can ensure that $PA(I - A)^{-1}f$ can be determined, though the exact solution of (20.1) is unknown.

Indeed, we construct the subspace H_P and operator P as follows. Let H_P be the subspace spanned by k linearly independent elements of some system $\{\varphi_i\}$ complete in H. The projection of an element $x \in H$ will be sought in the form

$$Px = \sum_{i=1}^{k} C_i\varphi_i. \tag{20.4}$$

The constants C_i will be determined so that the element $x - \Sigma_1^k C_i\varphi_i$ is orthogonal to the system of k linearly independent elements $B^*\psi_i$, where $\{\psi_i\}$ is a complete system in H, $B = I - A$, and B^* is the adjoint of B. We thus obtain a system of k linear algebraic equations in the k unknowns C_i:

$$\sum_{i=1}^{k} (\varphi_i, B^*\psi_j)\, C_i = (x, B^*\psi_j) \tag{20.5}$$

or

$$\sum_{i=1}^{k} (B\varphi_i, \psi_j)\, C_i = (Bx, \psi_j). \tag{20.6}$$

If the determinant of system (20.6) does not vanish, the elements C_i are uniquely determined. The element $PA(I - A)^{-1}f$ is also uniquely defined:

$$PA(I - A)^{-1}f = \sum_{i=1}^{k} C_i'\varphi_i,$$

where $\{C_i'\}$ is a solution of the system

$$\sum_{i=1}^{k} (B\varphi_i, \psi_j)\, C_i = (Af, \psi_j). \tag{20.7}$$

To solve equation (20.3), we use the ordinary method of successive approximations, defining the latter, which are of course also approximations to the solution of the equivalent equation (20.1), through the formula

$$x_n = f + PA(I - A)^{-1}f + QAx_{n-1}, \tag{20.8}$$

letting the initial approximation x_0 be an arbitrary element $x_0 \in H$.

It is readily seen that with the above choice of a projection P this algorithm is equivalent to an algorithm in which the successive approximations to the solution of (20.1) are defined by

$$x_n = f + Ax_{n-1} + \sum_{i=1}^{k} a_i\varphi_i \quad (x_0 \in H), \tag{20.9}$$

where the a_i are determined subject to the condition that $x_n - x$ (where x is the exact solution of (20.1)) is orthogonal to the elements $B^*\psi_i$ $(i = 1, \ldots, k)$.

Indeed, we can write (20.8) as

$$x_n = f + Ax_{n-1} + P[A(I - A)^{-1}f - Ax_{n-1}].$$

Thus the constants a_i in (20.9) and the constants c_i in

$$\sum_{i=1}^{k} c_i\varphi_i = P[A(I - A)^{-1}f - Ax_{n-1}]$$

are determined by the same system of equations.

If we set $\varphi_i = B^*\psi_i$ and $x_0 = \theta$ (where θ is the zero element), the first approximation of our algorithm will be the approximation obtained by the method of [55] with $x_0 = \theta$. Moreover, the determinant of the resulting system of equations will not vanish, since it is the Gram determinant of a system of linearly independent elements. The constants a_i in (20.9) are then determined from the condition that they minimize the norm $\|x_n - x\|$.

We now present some convergence criteria and error estimates for the process (20.8).

THEOREM 20.1. *Suppose that for some P the operator series*

$$I + L + L^2 + \ldots, \quad L = QAQ, \tag{20.10}$$

is convergent. Then the sequence $\{x_n\}$ converges to a solution x of equation (20.1) for any $x_0 \in H$, and the following error estimate holds:

$$\|x - x_n\| \leqslant \sigma_{n-p}\|QA\delta_p\| \quad (1 \leqslant p \leqslant n), \tag{20.11}$$

where

$$\sigma_{n-p} = \left\| \sum_{i=n-p}^{\infty} L^i \right\|; \quad \delta_p = x_p - x_{p-1}.$$

The proof is analogous to the convergence proof for the classical successive approximation method [235].

REMARK 1. If $\|L^q\| < 1$ for some q, then the series (20.10) is convergent, and then

$$\sigma_{n-p} \leqslant \frac{\| R \| \| L^{n-p} \|}{1 - \| L^q \|},$$

where $R = I + L + \cdots + L^{q-1}$.

We now assume that $P = P_k$ is a projection operator mapping H onto the k-dimensional subspace H_k defined above.

THEOREM 20.2. *If A is a compact operator and $P_k \longrightarrow I$ on H, i.e., $\lim P_k x = x$, then $\|L^q\| < 1$ for sufficiently large k, so that the successive approximations* (20.8) *converge to the solution of equation* (20.1).

PROOF. It suffices to observe that in this case $\|Q_k A\| \longrightarrow 0$ as $k \longrightarrow \infty$ ($Q_k = I - P_k$).

We now consider the case that $\{\varphi_i\}$ and $\{\psi_i\}$ are so chosen that P is an orthogonal projection: $P = P^2$ and P is selfadjoint. We then have the following two theorems.

THEOREM 20.3. *If $\|A\| < 1$, then the successive approximations converge and $\|L\| \leqslant \|A\|$.*

PROOF. We merely note that in this case $\|Q\| = 1$, and so $\|QAQ\| \leqslant A$.

THEOREM 20.4. *If $P_{k+m} = P_k + P_m$ and the convergence criterion of algorithm* (20.8) *holds for $P = P_k$, i.e.,*

$$\| L_k \| < 1, \; L_k = Q_k A Q_k, \; Q_k = I - P_k, \tag{20.12}$$

then the algorithm is convergent for $P = P_{k+m}$ as well, and $\|L_{k+m}\| \leqslant \|L_k\|$.

PROOF. This follows from the fact that $\|P_m\| = 1$.

Algorithm (20.8) may be generalized. Instead of (20.3), let us consider the more general equation

$$x = f + \bar{A}(1 - A)^{-1} f + (A - \bar{A}) x, \tag{20.13}$$

equivalent to equation (20.1) (where $\bar{A}$ is some operator close to A), to which we apply the classical method of successive approximations. If we choose $\bar{A}$ to have the form $C(I - A)$, where C is some operator, then the element $\bar{A}(I - A)^{-1} f$ is defined and $\bar{A}(I - A)^{-1} f = Cf$.

For example, we might define $\bar{A} = P^{(1)} A P^{(2)}$, where the $P^{(i)}$ are projections and $P^{(2)}$ is defined in the same way as P in equation (20.3). When $P^{(2)} = I$ and $P^{(1)} = P$, we get algorithm (20.8).

If $P^{(1)} = I$ and $P^{(2)} = P$, the above theorems will again hold, except that instead of the error estimate (20.11) we have

$$\| x - x_n \| \leqslant \| AQ \| \sigma_{n-p} \| Q \delta_p \| \qquad (1 \leqslant p \leqslant n). \tag{20.14}$$

REMARK 2. Let $\bar{A}$ be the operator

$$\bar{A} = PA = (I - P^*A)^{-1}P^*(I - A)A,$$

where P^* is some projection of H onto a subspace. We then obtain the method of averaged functional corrections, as considered in [121].

The operator $P = (I - P^*A)^{-1}P^*(I - A)$ is a projection, since

$$\begin{aligned} P^2 &= (I - P^*A)^{-1}P^*(I - A)(I - P^*A)^{-1}P^*(I - A) \\ &= (I - P^*A)^{-1}P^*(I - P^*A)(I - P^*A)^{-1}P^*(I - A) \\ &= (I - P^*A)^{-1}P^*(I - A) = P. \end{aligned}$$

§21. Solution of some types of operator equation in a semiordered Hilbert space

In this section, by constructing a special sequence of successive approximations, we shall establish existence and uniqueness theorems for the operator equation

$$x = \varphi + Afx, \tag{21.1}$$

where A is a selfadjoint linear operator in a semiordered Hilbert space H, f is a nonlinear operator defined in H, and φ is a known element of H.

1. Let us assume that the operator f has a Fréchet derivative, and that

$$g \leqslant f'(x) \leqslant h, \quad x \in H, \tag{21.2}$$

where g and h are selfadjoint operators. Set $\mu = (g + h)/2$. Let $\mu \geqslant 0$ or $\mu \leqslant 0$, and $\bar{\mu} = \mu$ if $\mu > 0$, $\bar{\mu} = -\mu$ if $\mu < 0$.

THEOREM 21.1. *If*

$$\|h - g\| \|(h + g)^{-1}\| < \rho(1, X_{\bar{A}}), \tag{21.3}$$

where $\rho(1, X_{\bar{A}})$ denotes the distance from 1 to the spectrum $X_{\bar{A}}$ of the operator $\bar{A} = \sqrt{\bar{\mu}}A\sqrt{\bar{\mu}}$, then equation (21.1) *has a unique solution x in H, which is the limit of the successive approximations x_n defined by*

$$x_n = \varphi + A(f - \bar{\mu})x_{n-1} + A\bar{\mu}x_n \quad (n = 1, 2, \ldots) \tag{21.4}$$

(x_0 is an arbitrary element of H). Moreover, the following error estimate is valid:

$$\|x - x_n\| \leqslant \|(\sqrt{\bar{\mu}})^{-1}\|(1 - \bar{q}_T)^{-1}\bar{q}_T^{n-p+1}\|y_p - y_{p-1}\| \quad (1 \leqslant p \leqslant n), \tag{21.5}$$

where $y_n = \sqrt{\bar{\mu}}x_n$ and $\bar{q}_T = \|h - g\|\|(h + g)^{-1}\|\rho^{-1}(1, X_{\bar{A}})$.

PROOF. It is clear that equation (21.1) is equivalent to the equation $x = \varphi + A(f - \bar{\mu})x + A\bar{\mu}x$. For $y = \sqrt{\bar{\mu}}x$, we have

$$y = \sqrt{\bar{\mu}}\,\varphi + \sqrt{\bar{\mu}}\,A\,[f(\sqrt{\bar{\mu}})^{-1} - \sqrt{\bar{\mu}}\,]\,y + \sqrt{\bar{\mu}}\,A\sqrt{\bar{\mu}}\,y,$$

whence

$$y = (I - \bar{A})^{-1}\sqrt{\bar{\mu}}\,\varphi + (I - \bar{A})^{-1}\bar{A}\,\bar{\mu}^{-1}\,[\sqrt{\bar{\mu}}\,f(\sqrt{\bar{\mu}})^{-1} - \bar{\mu}]\,y. \quad (21.6)$$

Now the operator $\bar{A}$ is by assumption selfadjoint, and thus the norm of the operator $(I - \bar{A})^{-1}\bar{A}$ is

$$\|(I - \bar{A})^{-1}\bar{A}\| = \sup_{t \in X_{\bar{A}}} |t - 1|^{-1} = \rho^{-1}(1, X_{\bar{A}}). \quad (21.7)$$

We thus have an upper bound $\bar{q}_T$ for the Lipschitz operator q_T of the operator T defined by the right-hand side of (21.6), $q_T \leqslant \bar{q}_T$. Consequently, we can apply Banach's contraction mapping principle to equation (21.6), so that the successive approximations $y_n = Ty_{n-1}$, $y_0 \in H$, converge to a solution of (21.6).

The sequence $\{x_n\}$ will then converge to a solution of (21.1). The error estimate (21.5) is established in the usual way.

If A is the identity operator ($A = I$) and $g = mI$ and $h = MI$, where m and M are numbers, we obtain the corresponding results of [175] and [177]. The case in which equation (21.1) is a nonlinear integral equation of Hammerstein type is investigated in [45] and [230] under slightly different assumptions.

2. Let H be the Hilbert space of sequences $x = \{x_i\}$ ($i = 1, \ldots, k$; k is either a natural number or infinity), where x_i are elements of Hilbert spaces H_i, with the scalar product defined by $(x, y) = \Sigma_1^k (x_i, y_i)_{H_i}$ (the symbol $(x_i, y_i)_{H_i}$ denotes the scalar product in H_i). Assuming that $\Sigma_1^k \|x_i\|_{H_i}$ is convergent for each $x \in H$, we define the norm in H, as usual, by $\|x\| = \sqrt{\Sigma_1^k \|x_i\|_{H_i}^2}$.

Let A be a selfadjoint operator, defined through a matrix whose entries are operators A_{ij} ($i, j = 1, \ldots, k$), i.e. $A = \{A_{ij}\}$, where each A_{ij} is an operator from H_j to H_i. Let f be an operator $fx = \{f_i x_i\}$, where f_i is an operator from H_i to H_i. Then equation (21.1) may be expressed as a system of equations

$$x_i = \varphi_i + \sum_{j=1}^{k} A_{ij} f_j x_j \quad (i = 1, 2, \ldots, k). \quad (21.8)$$

We shall assume that each of the operators f_i has a Fréchet derivative satisfying the condition

$$g_i \leqslant f_i'(x_i) \leqslant h_i, \quad (21.9)$$

where g_i and h_i are selfadjoint operators in H_i. Thus,

$$g \leqslant f'(x) \leqslant h, \; g = \{g_i\}, \; h = \{h_i\}. \quad (21.10)$$

We have the following estimate for the Lipschitz operator q of the operator

$$\sqrt{\bar{\mu}}f\,(\sqrt{\bar{\mu}})^{-1} - \bar{\mu} = \{\sqrt{\bar{\mu}_i}f_i(\sqrt{\bar{\mu}_i})^{-1} - \bar{\mu}_i\},$$

wnere $\bar{\mu}_i = \mu_i$ if $\mu_i > 0$, $\bar{\mu}_i = -\mu_i$ if $\mu_i < 0$, and $\mu_i = (g_i + h_i)/2$:

$$q \leqslant \frac{1}{2} \| h - g \| \leqslant \frac{1}{2} \sup_{i=1,\ldots,k} \| h_i - g_i \|_{H_i}, \tag{21.11}$$

and the norm of the operator $(g + h)^{-1}$ satisfies the estimate

$$\| (g + h)^{-1} \| \leqslant \sup_{i=1,\ldots,k} \| (g_i + h_i)^{-1} \|_{H_i}. \tag{21.12}$$

Then the operator $\bar{A}$ is defined by $\bar{A} = \{\sqrt{\bar{\mu}_i} A_{ij} \sqrt{\bar{\mu}_j}\}$, and the iterative process may be written

$$x_{i,n} = \varphi_i + \sum_{j=1}^{k} A_{ij} (f_j - \mu_j) x_{j,n-1} + \sum_{j=1}^{k} A_{ij} \mu_j x_{j,n} \quad (n = 1, 2, \ldots). \tag{21.13}$$

REMARK 1. If each H_i is a function space, we may replace (21.9) by

$$g_i \leqslant [f_i (x_i) - f_i (y_i)] (x_i - y_i)^{-1} \leqslant h_i, \tag{21.14}$$

where g_i and h_i are suitable functions.

3. Consider the system of nonlinear operator equations

$$x_i = \varphi_i + A_i f_i (x_1, x_2, \ldots, x_k) \quad (i = 1, 2, \ldots, k), \tag{21.15}$$

where each operator A_i has domain and range in H_i and each operator f_i maps H into H_i (see subsection 2). Setting $A = \{A_i\}$, $f = \{f_i\}$ and $\varphi = \{\varphi_i\}$, we can write the system (21.15) in the form (21.1).

Let A_i be selfadjoint operators, and assume that the f_i have Fréchet derivatives $f_i' = \{f_{ij}'\}$ satisfying the condition

$$g_i \leqslant f_{ij} (x_1, \ldots, x_k) \leqslant h_i \tag{21.16}$$

(where f_{ij} denotes the partial derivative of $f_i(x_1, \ldots, x_k)$ with respect to x_j), where $\mu_i = (g_i + h_i)/2$ are selfadjoint operators in H_i, $\mu_i \geqslant 0$ or $\mu_i \leqslant 0$, $\bar{\mu}_i = \mu_i$ if $\mu_i > 0$ and $\bar{\mu}_i = -\mu_i$ if $\mu_i \leqslant 0$. Suppose that the f_i' s satisfy the condition

$$\| f_i x - f_i y \|_{H_i} \leqslant \sum_{j=1}^{k} c_{ij} \| x_j - y_j \|_{H_j} \quad (x_j, y_j \in H_j). \tag{21.17}$$

Setting $T = \{T_i\}$, where

$$T_i x = (I - \bar{A}_i)^{-1} \bar{A}_i \bar{\mu}_i^{-1} \left\{ \sqrt{\bar{\mu}_i} f_i \left[\left(\sqrt{\bar{\mu}_1}\right)^{-1} x_1, \ \ldots, \ \left(\sqrt{\bar{\mu}_k}\right)^{-1} x_k \right] - \bar{\mu}_i x_i \right\},$$
$$\bar{A}_i = \sqrt{\bar{\mu}_i} A_i \sqrt{\bar{\mu}_i}, \tag{21.18}$$

we readily see that the Lipschitz constant for this operator admits an estimate

$$q_T^2 \leqslant \bar{q}_T^2 = \sum_{i=1}^{k} \sup_{t_i \in X_{\bar{A}_i}} | t_i - 1 |^{-2} \Big[\| h_i - g_i \|_{H_i}^2 + \sum_{j=1}^{k}{}' \| \sqrt{\bar{\mu}_i} \|_{H_i}^2 c_{ij}^2 \| (\sqrt{\bar{\mu}_j})^{-1} \|_{H_j}^2 \Big] \| (g_i + h)^{-1} \|_{H_i}^2 \tag{21.19}$$

(the prime indicates summation over $j \neq i$).

Hence we deduce the following theorem:

THEOREM 21.2 *If* $\bar{q}_T < 1$, *the system of equations* (21.15) *has a unique solution* $x \in H$, *which is the limit of the iterative process defined by the system of equations*

$$x_{i,n} = \varphi_i + A_i\,[f_i\,(x_{1,n-1}, \ldots, x_{k,n-1}) - \mu_i x_{i,n-1}] + A_i \mu_i x_{i,n} \qquad (21.20)$$

($i = 1, 2, \ldots, k$; $n = 1, 2, \ldots$). *Error estimates for this process are given by*

$$\| x - x_n \| \leqslant (1 - \bar{q}_T)^{-1} \bar{q}_T^{\,n-p+1} \| \mu^{-1} \| \| y_p - y_{p-1} \| \quad (1 \leqslant p \leqslant n), \qquad (21.21)$$

$$\| x_i - x_{i,n} \|_{H_i} \leqslant \bar{q}_{T_i} \| \mu_i^{-1} \|_{H_i} (1 - \bar{q}_T)^{-1} \bar{q}_T^{\,n-p} \| y_p - y_{p-1} \| \quad (1 \leqslant p \leqslant n), \qquad (21.22)$$

where $y_{i,n} = \sqrt{\bar{\mu}_i}\, x_{i,n}$ *and*

$$\bar{q}_{T_i}^2 = \sup_{t_i \in X_{\bar{A}_i}} | t_i - 1 |^{-2} \Big[\| h_i - g_i \|_{H_i}^2 + \sum_{j=1}^{k}{}' \| \sqrt{\bar{\mu}_i} \|_{H_i}^2 c_{ij}^2 \| (\sqrt{\bar{\mu}_j})^{-1} \|_{H_j}^2 \Big] \| (g_i + h_i)^{-1} \|_{H_i}^2.$$

REMARK 2. If we replace the Hilbert space by a space normed by $\|x\| = \sup_{1 \leqslant i \leqslant k} \|x_i\|_{H_i}$, the analog of Theorem 21.2 is valid, except that $\bar{q}_T$ and $\bar{q}_{T_i}$ must be replaced by the quantities $\bar{\bar{q}}_T$ and $\bar{\bar{q}}_{T_i}$ defined by

$$\bar{\bar{q}}_T = \sup_{i=1,\ldots,k} \sup_{t_i \in X_{\bar{A}_i}} | t_i - 1 |^{-1} \Big[\| h_i - g_i \|_{H_i} + \sum_{j=1}^{k}{}' \| \mu_i \|_{H_i} c_{ij} \| (\sqrt{\bar{\mu}_j})^{-1} \|_{H_j} \Big] \| (g_i + h_i)^{-1} \|_{H_i},$$

$$\bar{\bar{q}}_{T_i} = \sup_{t_i \in X_{\bar{A}_i}} | t_i - 1 |^{-1} \Big[\| h_i - g_i \|_{H_i} + \sum_{j=1}^{k}{}' \| \sqrt{\bar{\mu}_i} \|_{H_i} c_{ij} \| (\sqrt{\bar{\mu}_j})^{-1} \|_{H_j} \Big] \| (g_i + h_i)^{-1} \|_{H_i}.$$

§22. Direct-product Banach spaces with *m*-norm and projections P_N. Error estimates for the projection method

In this section we consider a special Banach space, which is a generalization of the space of bounded number sequences. We shall also study certain projection operators and their properties.

Let E be a space whose elements are sequences $x = \{x_i\}$ ($i = 1, \ldots, m$), where the x_i are elements of Banach spaces B_i, and m is a natural number or infinity. We shall assume that the sequences $x = \{x_i\}$ are bounded: there exists a constant c such that $\|x_i\|_{B_i} \leqslant c < \infty$ (where $\|x_i\|_{B_i}$ denotes the norm of x_i in B_i).

Addition and multiplication by scalars are defined as in ordinary vector spaces: if $x = \{x_i\}$ and $y = \{y_i\}$, then

$$x + y = \{x_i + y_i\}, \tag{22.1}$$

$$\alpha x = \{\alpha x_i\}. \tag{22.2}$$

The norm of an element $x \in E$ is defined by

$$\| x \| = \sup_i \| x_i \|_{B_i}. \tag{22.3}$$

With this definition, E is a Banach space, which we call a direct-product Banach space.

Define a projection P_N ($P_N = P_N^2$) by

$$P_N x = \{x_i^{(P)}\}, \text{ where } x_i^{(P)} = \begin{cases} x_i & \text{for } i = 1, 2, \ldots, N, \\ 0 & \text{for } i = N + 1, \ldots, m, \end{cases} \tag{22.4}$$

where N is a natural number: if E is the space of finite m-termed sequences, we assume of course that $N \leqslant m$.

It is obvious that the operator $Q_N = I - P_N$ (where I is the identity operator) is also a projection ($Q_N = Q_N^2$) and

$$Q_N x = \{x_i^{(Q)}\}, \text{ where } x_i^{(Q)} = \begin{cases} 0 & \text{for } i = 1, \ldots, N, \\ x_i & \text{for } i = N + 1, \ldots, m. \end{cases} \tag{22.5}$$

It is easily shown that the projections P_N and Q_N thus defined satisfy the condition that, for any $x, y \in E$,

$$\| P_N x + Q_N y \| = \max \{\| P_N x \|, \| Q_N y \|\}. \tag{22.6}$$

In fact,

$$P_N x + Q_N y = \{z_i\}; \; z_i = \begin{cases} x_i & \text{for } i = 1, \ldots, N, \\ y_i & \text{for } i = N + 1, \ldots, m, \end{cases} \tag{22.7}$$

$$\begin{aligned} \| P_N x + Q_N y \| = \sup_i \| z_i \|_{B_i} &= \max \{ \sup_{i=1,\ldots,N} \| x_i \|_{B_i}, \sup_{i=N+1,\ldots,m} \| y_i \|_{B_i} \} \\ &= \max \{\| P_N x \|, \| Q_N y \|\}, \end{aligned} \tag{22.8}$$

as required.

Property (22.6) will enable us to derive new convergence criteria, which do not follow from the general theory of Chapter I, and also to obtain new and more accurate error estimates both for the projection method and for some iterative projection methods.

The operator equation $x = Tx$, where T is an operator in an m-normed direct-product Banach space E, may be solved by the projection method using the projection P_N. In other words, we replace the equation $x = Tx$ by the "approximating" equation

$$y = P_N T y, \tag{22.9}$$

whose solution is then treated as an approximate solution of the original equation.

Using property (22.6) of the projections P_N and Q_N, we prove the following result.

THEOREM 22.1. *Suppose that in some subset D of E the equation $x = Tx$ has a solution x and equation (22.9) a solution y, and that*

$$q_{PT} < 1, \tag{22.10}$$

where q_{PT} is a Lipschitz constant for $P_N T$ in D. Then the following error estimate holds:

$$\|x - y\| \leqslant (1 - q_{QT})^{-1} \|Q_N T y\|, \tag{22.11}$$

where q_{QT} is a Lipschitz constant $Q_N T$ in D.

PROOF. Since x and y are solutions of the equations $x = Tx$ and (22.9), respectively, we have

$$x - y = Tx - P_N Ty + P_N Tx - P_N Tx = Q_N Tx + P_N Tx - P_N Ty.$$

Hence, by property (22.6),

$$\|x - y\| = \max \{\|Q_N Tx\|, \|P_N Tx - P_N Ty\|\}. \tag{22.12}$$

Since the equality

$$\|x - y\| = \|P_N Tx - P_N Ty\| \tag{22.13}$$

cannot be true, because it implies the inequality

$$\|x - y\| \leqslant q_{PT} \|x - y\|,$$

which is impossible when $0 < q_{PT} < 1$, $\|x - y\| \neq 0$, it follows that

$$\|x - y\| = \|Q_N Tx\|. \tag{22.14}$$

But then

$$\|x - y\| \leqslant \|Q_N Tx - Q_N Ty\| + \|O_N Ty\| \leqslant q_{QT} \|x - y\| + \|Q_N Ty\|,$$

and this implies the required estimate (22.11).

THEOREM 22.2. *If T is a compact operator, mapping a ball $S(P_N g, r)$, where g is a fixed element of E, into itself (i.e., $\|Tx - P_N g\| \leqslant r$ for $x \in S$), and condition (22.10) holds for the ball S, then each of the equations $x = Tx$ and (22.9) is solvable in S, and an error estimate is given by (22.11), where q_{QT} corresponds to S.*

PROOF. We need only show that the equation (22.9) is solvable in S, for the solvability of $x = Tx$ in S follows from the Schauder principle. But this is also obvious: for any $x \in E$ we have

$$\| P_N x \| \leqslant \| x \|. \tag{22.15}$$

Thus

$$\| P_N Tx - P_N g \| = \| P_N (Tx - P_N g) \| \leqslant \| Tx - P_N g \| \leqslant r \tag{22.16}$$

for any $x \in S$. And this means that $P_N T$ maps S into itself, and so, again by the Schauder principle, equation (22.9) is solvable in S.

The rest of the proof follows directly from that of Theorem 22.1.

In conclusion, we note that (22.11) is generally sharper than the estimate furnished by the general comparison theorem of [169], when the operator S in the latter approximating T is defined as $P_N T$, the Lipschitz operator reduces to a Lipschitz constant, and the equation is considered in the Banach space E.

§23. Generalized iterative projection method for operator equations in an m-normed direct-product Banach space with projection operator P_N

We shall now derive simple convergence criteria for the generalized iterative projection method of §§8, 9 and 14, for operator equations in more general spaces. Recall that the idea of the method is to define the successive approximations x_n to the solution of $x = Tx$ as solutions of the equations

$$x_n = F_k (x_n, x_{n-1}) \ (x_0 \in E, \ n = 1, 2, \ldots), \tag{23.1}$$

where the F_k are defined by

$$\begin{aligned} F_1 (x, y) &= F (x, y), \\ F_i (x, y) &= F [x, F_{i-1} (x, y)] \quad (i = 2, 3, \ldots), \end{aligned} \tag{23.2}$$

and $F(x, y)$ is defined in one of the following ways:

a) $F (x, y) = P_N Tx + Q_N Ty$;

b) $F (x, y) = T (P_N x + Q_N y)$;

c) $F (x, y) = P_N Tx + Q_N T (P_N x + Q_N y)$, $Q_N = I - P_N$.

THEOREM 23.1. *Let T be an operator in an m-normed space E, satisfying a Lipschitz condition with constant $q_T < 1$. Then each of equations* (23.1) *is uniquely solvable in E for x_n, in each of the three cases* (a), (b), (c), *and the sequence $\{x_n\}$ of solutions converges to a solution of the equation $x = Tx$, with the following error estimates*:

Case (a):

$$\| x - x_n \| \leqslant \frac{q_{Q_N T}^{k(n-p+1)}}{1 - q_{Q_N T}^k} \| \delta_p \| \quad (1 \leqslant p \leqslant n). \tag{23.3}$$

Case (b):

$$\|x - x_n\| \leqslant q_T \frac{q_{Q_NT}^{k(n-p+1)-1}}{1 - q_{Q_NT}^k} \|Q_N\delta_p\| \quad (1 \leqslant p \leqslant n). \tag{23.4}$$

Case (c):

$$\|x - x_n\| \leqslant \frac{q_{Q_NT}^{k(n-p+1)}}{1 - q_{Q_NT}^k} \|Q_N\delta_p\| \quad (1 \leqslant p \leqslant n). \tag{23.5}$$

Here $\delta_p = x_p - x_{p-1}$.

PROOF. We first consider case (a). Since

$$F_1(x_n, x_{n-1}) = P_NTx_n + Q_NTx_{n-1},$$
$$F_i(x_n, x_{n-1}) = P_NTx_n + Q_NTF_{i-1}(x_n, x_{n-1}) \quad (i = 2, 3, \ldots),$$

it follows from (23.1) via (22.6) that

$$\|\delta_n\| = \max\{\|P_NTx_n - P_NTx_{n-1}\|, \|Q_NTF_{k-1}(x_n, x_{n-1}) - Q_NTF_{k-1}(x_{n-1}, x_{n-2})\|\}.$$

But the equality

$$\|\delta_n\| = \|P_NTx_n - P_NTx_{n-1}\| \tag{23.6}$$

is impossible, for $q_T < 1$ implies $q_{P_NT} \leqslant q_T < 1$ and $\|\delta_n\| \leqslant q_{P_NT}\|\delta_n\|$, and when $q_{P_NT} < 1$ this can be true only if $\|\delta_n\| = 0$. Thus, we have

$$\|\delta_n\| = \|Q_NTF_{k-1}(x_n, x_{n-1}) - Q_NTF_{k-1}(x_{n-1}, x_{n-2})\|. \tag{23.7}$$

It follows that

$$\|\delta_n\| \leqslant q_{Q_NT} \max\{\|P_NTx_n - P_NTx_{n-1}\|, \|Q_NTF_{k-2}(x_n, x_{n-1}) - Q_NTF_{k-2}(x_{n-1}, x_{n-2})\|\},$$

whence

$$\|\delta_n\| \leqslant q_{Q_NT}\|Q_NTF_{k-2}(x_n, x_{n-1}) - Q_NTF_{k-2}(x_{n-1}, x_{n-2})\|.$$

Repeating this procedure for the right-hand side, we finally obtain

$$\|\delta_n\| \leqslant q_{Q_NT}^k \|\delta_{n-1}\|, \tag{23.8}$$

and this readily implies the desired conclusion.

In case (b), we note that

$$F_1(x_n, x_{n-1}) = T(P_Nx_n + Q_Nx_{n-1}),$$
$$F_i(x_n, x_{n-1}) = T[P_Nx_n + Q_NF_{i-1}(x_n, x_{n-1})] \quad (i = 2, 3, \ldots),$$

and use property (22.6) to deduce from (23.1) that

$$\|\delta_n\| \leqslant q_T \max\{\|P_N\delta_n\|, \|Q_NT_{k-1}(x_n, x_{n-1}) - Q_NT_{k-1}(x_{n-1}, x_{n-2})\|\},$$

whence it follows that

$$\|\delta_n\| \leqslant q_T \| Q_N T_{k-1}(x_n, x_{n-1}) - Q_N T_{k-1}(x_{n-1}, x_{n-2}) \|, \tag{23.9}$$

since $q_T < 1$ implies that $\|\delta_n\| \leqslant q_T \|P_N \delta_n\| \leqslant q_T \|\delta_n\|$ cannot hold except in the trivial cases $\|\delta_n\| = 0$ or $q_T = 0$.

Repeating the procedure for the right-hand side of (23.9), we finally obtain

$$\|\delta_n\| \leqslant q_T q_{Q_N T}^{k-1} \| Q_N \delta_{n-1} \|. \tag{23.10}$$

As before, we now deduce from (23.1) that

$$\| Q_N \delta_n \| \leqslant q_{Q_N T}^k \| Q_N \delta_{n-1} \|. \tag{23.11}$$

Inequalities (23.10) and (23.11) now readily imply the desired conclusion.

Finally, consider case (c). We have

$$F_1(x_n, x_{n-1}) = P_N T x_n + Q_N T (P_N x_n + Q_N x_{n-1}),$$
$$F_i(x_n, x_{n-1}) = P_N T x_n + Q_N T [P_N x_n + Q_N F_{i-1}(x_n, x_{n-1})]$$
$$(i = 2, 3, \ldots).$$

Using (22.6) and proceeding as in the previous cases, we see from (23.1) that

$$\|\delta_n\| \leqslant q_{Q_N T}^k \| Q_N \delta_{n-1} \|, \tag{23.12}$$

$$\| Q_N \delta_n \| \leqslant q_{Q_N T}^k \| Q_N \delta_{n-1} \|, \tag{23.13}$$

and hence the conclusion of the theorem.

No difficulties are involved in proving the analog of Theorem 23.1 for the case that the conditions hold in a subset of E rather than throughout the space. For example, let us consider case (a).

Suppose that in a set D defined by the inequalities $\|x_i - g_i\|_{B_i} \leqslant r_i$, where $g = \{g_i\}$ is a fixed element of E and r_i is to be determined, we have

$$\| T_i x - g_i \|_{B_i} \leqslant F^{(i)}(r_1, \ldots, r_m) \quad (i = 1, 2, \ldots, N),$$
$$\| T_i y - g_i \|_{B_i} \leqslant F^{(i)}(r_1, \ldots, r_m) \quad (i = N+1, \ldots, m).$$

If the system of inequalities

$$F^{(i)}(r_1, \ldots, r_m) \leqslant r_i \quad (i = 1, 2, \ldots, m) \tag{23.14}$$

has a solution $r_i = r_i^* > 0$ and the assumptions of Theorem 23.1 hold in D^* ($\|x_i - g_i\|_{B_i} \leqslant r_i^*$), then the conclusion of the theorem is also valid in D^*.

§24. Modified iterative projection method for operator equations in an m-normed direct-product Banach space with projection operator P_N

In this section we present some simple convergence criteria and error estimates for the algorithm of §§11, 12 and 15, for operator equations in more general spaces.

Applied to the equation $x = Tx$, this algorithm dictates that we define the successive approximations as $x_n = u_n + v_n$, where u_n, v_n is a solution of the system of operator equations

$$u_n = P_N T(u_n + v_n), \qquad v_n = Q_N S_k(u_n, v_{n-1}), \tag{24.1}$$

where the S_k are defined by

$$\begin{aligned} S_1(u, v) &= Q_N T(u + v), \\ S_i(u, v) &= Q_N T[u + S_{i-1}(u, v)] \quad (i = 2, 3, \ldots). \end{aligned} \tag{24.2}$$

THEOREM 24.1. *Let T be an operator in an m-normed space, satisfying a Lipschitz condition with constant $q_T < 1$. Then the system of operator equations (24.1) is uniquely solvable for u_n and v_n, and the sequence $\{x_n\}$, $x_n = u_n + v_n$, converges to a solution of the equation $x = Tx$, with error estimate*

$$\|x - x_n\| \leqslant \frac{q_{Q_N T}^{k(n-p+1)}}{1 - q_{Q_N T}^k} \|\delta v_p\| \quad (1 \leqslant p \leqslant n), \tag{24.3}$$

where $\delta v_p = v_p - v_{p-1}$.

PROOF. Substituting the expression for v_n from the second equation of (24.1) into the first, we get an equation for Pu_n:

$$Pu_n = P_N T[Pu_n + Q_N S_k(Pu_n, v_{n-1})]. \tag{24.4}$$

It is readily seen that because of property (22.6) and the inequality $q_T < 1$ the operator on the right of (24.4) satisfies a Lipschitz condition with constant less than 1. Thus, by the contraction mapping principle, equation (24.4) has a unique solution Pu_n. Substituting this solution into the right-hand side of the second equation of (24.1), we find v_n. Thus u_n and v_n are uniquely determined by (24.1).

Since obviously $u_n = P_N u_n$ and $v_n = Q_N v_n$, it follows from the first equation of (24.1) that

$$\|P\delta u_n\| \leqslant q_{P_N T} \|Q\delta v_n\|, \tag{24.5}$$

since the inequality $\|P_N \delta u_n\| \leqslant q_{P_N T} \|P_N \delta u_n\|$ cannot hold with $q_T < 1$ unless $\|P_N \delta u_n\| = 0$.

From the second equation of (24.1), we deduce

$$\|Q_N \delta v_n\| \leqslant q_{Q_N T} \max\{\|P\delta u_n\|, \|Q_N S_{k-1}(u_n, v_{n-1}) - Q_N S_{k-1}(u_{n-1}, v_{n-2})\|\},$$

whence

$$\|Q_N \delta v_n\| \leqslant \|Q_N S_{k-1}(u_n, v_{n-1}) - Q_N S_{k-1}(u_{n-1}, v_{n-2})\|,$$

since the inequality $\|Q_N \delta v_n\| \leqslant q_{Q_N T} \|P\delta u_n\|$ is impossible. Indeed, as $0 < q_{Q_N T} < 1$, substitution of this inequality into the right-hand side of (24.5) would yield

$$\| P_N \delta u_n \| \leqslant q_{P_N T}\, q_{Q_N T} \| P\delta u_n \|.$$

One shows in analogous fashion that, in general,

$$\| Q_N \delta v_n \| \leqslant q^k_{Q_N T} \| Q_N \delta v_{n-1} \|. \tag{24.6}$$

It follows that $\{v_n\}$ is a Cauchy sequence. Since $x_n = P_N T x_n + Q_n v_n$, it follows that $\|\delta x_n\| \leqslant \|Q_N \delta v_n\|$, and so $\{x_n\}$ is also a Cauchy sequence.

As before, one shows that the sequences u_n and v_n converge to a solution u, v of the system

$$u = P_N T (u + v), \qquad v = Q_N T (u + v) \tag{24.7}$$

and so $\{x_n\}$ converges to a solution of the equation $x = Tx$.

The error estimate (24.3) is established in the usual way.

§25. *m*-normed direct-product Banach space and projection *R*. Error estimate for projection method

Besides the projection P_N defined in (22.4), we consider another projection operator Π, defined by

$$\Pi x = \{P_i x_i\}, \tag{25.1}$$

where the P_i are projections of the spaces B_i, respectively.

Let R denote the product of P_N and Π: $R = P_N \Pi = \Pi P_N$. Obviously,

$$Rx = \{x_i^{(R)}\},\ x_i^{(R)} = \begin{cases} P_i x_i & (i = 1, 2, \ldots, N), \\ 0 & (i = N+1, \ldots, m). \end{cases} \tag{25.2}$$

The operators $Q_{P_N} = I - P_N$, $Q_\Pi = I - \Pi$ and $Q_R = I - R$ are of course also projections, and

$$Q_{P_N} x = \{\bar{x}_i^{(P)}\};\ \bar{x}_i^{(P)} = \begin{cases} 0 & (i = 1, 2, \ldots, N), \\ x_i & (i = N+1, 2, \ldots, m); \end{cases} \tag{25.3}$$

$$Q_\Pi x = \{Q_i x_i\}; \tag{25.4}$$

$$Q_R x = \{\bar{x}_i^{(R)}\};\ \bar{x}_i^{(R)} = \begin{cases} Q_i x_i & (i = 1, \ldots, N), \\ x_i & (i = N+1, \ldots, m). \end{cases} \tag{25.5}$$

It is easy to see that for any $x, y \in E$ (where E is an m-normed direct-product Banach space)

$$\| Rx + Q_R y \| = \max \{\| Rx + P_N Q_R y \|,\ \| Q_{P_N} y \|\}. \tag{25.6}$$

Indeed, by the definitions of the operators involved,

$$Rx + Q_R y = P_N \Pi x + P_N Q_R y + Q_{P_N} y,$$

whence, in view of the fact that (22.6) holds for any $x, y \in E$, we obtain (25.6).

If the B_i are Hilbert spaces and the P_i orthogonal projections ($P_i = P_i^2$ and P_i is selfadjoint), then for arbitrary $x, y \in E$

$$\| Rx + Q_R y \|^2 \leqslant \max \{\| Rx \|^2 + \| P_N Q_R y \|^2, \| Q_{P_N} y \|^2\}. \tag{25.7}$$

This follows at once from (25.6), since under these conditions we have

$$\| P_i x_i + Q_i y_i \|^2_{B_i} = \| P_i x_i \|^2_{B_i} + \| Q_i y_i \|^2_{B_i}. \tag{25.8}$$

Let us solve the equation $x = Tx$ in E, using the projection method with the projection R, thus replacing the equation by the "approximating" equation

$$y = RTy. \tag{25.9}$$

Using the property (25.6) of the operators R and Q_R, we shall prove the following theorem, which furnishes an error estimate.

THEOREM 25.1. *Suppose that in some subset G of E the equation $x = Tx$ has a solution x, and equation* (25.9) *a solution y. Then an error estimate is given by*

$$\| x - y \| \leqslant \max \{(1 - q_{Q_{P_N}T})^{-1} \| Q_{P_N} Ty \| \; (1 - q_{P_N T})^{-1} \| P_N Q_R Ty \|\}. \tag{25.10}$$

(Here, as usual, the letter q, suitably indexed, will denote a Lipschitz constant for the appropriate operator in the set G.)

PROOF. Since x and y are solutions of the equations $x = Tx$ and (25.9), it is clear that

$$x - y = Tx - RTy + RTx - RTx = RTx - RTy + Q_R Tx$$

and by property (25.6)

$$\| x - y \| = \max \{\| (RTx - RTy) + P_N Q_R Tx \|, \| Q_{P_N} Tx \|\}. \tag{25.11}$$

If $\|x - y\| = \|Q_{P_N} Tx\|$, then

$$\| x - y \| \leqslant \| Q_{P_N} Tx - Q_{P_N} Ty \| + \| Q_{P_N} Ty \| \leqslant q_{Q_{P_N}T} \| x - y \| + \| Q_{P_N} Ty \|,$$

whence

$$\| x - y \| \leqslant (1 - q_{Q_{P_N}T})^{-1} \| Q_{P_N} Tx \|. \tag{25.12}$$

But if $\|x - y\| = \|RTx - RTy + P_N Q_R Tx\|$, then

$$\begin{aligned} \| x - y \| &= \| RTx - RTy + P_N Q_R Tx - P_N Q_R Ty + P_N Q_R Ty \| \\ &= \| P_N Tx - P_N Ty + P_N Q_R Ty \| \leqslant q_{P_N T} \| x - y \| + \| P_N Q_R Ty \| \end{aligned}$$

and

$$\|x-y\| \leqslant (1-q_{P_N T})^{-1} \|P_N Q_R T y\|. \tag{25.13}$$

Comparing (25.12) and (25.13), we obtain the desired estimate (25.10).

Note that in some cases the estimate (25.10) is sharper than that following directly from (2.15), if the Lipschitz operator is replaced by a Lipschitz constant and the equation considered in the Banach space E.

§26. Investigation of generalized iterative projection methods for equations in an m-normed direct-product Banach space with projection R

We now present some convergence criteria and error estimates for the generalized iterative method defined by equations (23.1) and (23.2), in the following cases:

a) $F(x, y) = RTx + Q_R Ty$,

b) $F(x, y) = T(Rx + Q_R y)$

applied to the equation $x = Tx$ in an m-normed space E. The operators R and Q_R were defined in §25.

Using (25.6), which is true in the present case, we prove

THEOREM 26.1. *If*

$$\max\{r+p, q\} < 1, \tag{26.1}$$

where $r = q_{RT}$, $p = q_{P_N Q_R T}$ *and* $q = q_{Q_{P_N} T}$ *(the q's are Lipschitz constants for the appropriate operators), then the successive approximations* x_n *are uniquely determined by equations* (23.1) *in both cases* (a) *and* (b), *and converge to the unique solution in E of the equation* $x = Tx$. *The following error estimates hold*:

Case (a):

$$\|x-x_n\| \leqslant (1-\varepsilon_k)^{-1} \varepsilon_k^{n-p+1} \|x_p - x_{p-1}\| \quad (1 \leqslant p \leqslant n). \tag{26.2}$$

Case (b):

$$\|x-x_n\| \leqslant q_T (1-\varepsilon_k)^{-1} \varepsilon_k^{n-p+1} \|z_p - z_{p-1}\| \quad (1 \leqslant p \leqslant n). \tag{26.3}$$

Here the ϵ_i *are defined recursively by*

$$\varepsilon_i = \max\left\{\frac{p\varepsilon_{i-1}}{1-r\varepsilon_{i-1}},\ q\varepsilon_{i-1}\right\} < 1 \quad (\varepsilon_0 = 1) \tag{26.4}$$

and

$$z_n = Rv_n + Q_R S_{k-1}(Rv_n, Q_R v_{n-1}) \quad (v_0 = Tz_0),$$
$$S_1(Rx, Q_R y) = T(Rx + Q_R y), \tag{26.5}$$

$$S_i(Rx + Q_R y) = T[Rx + Q_R S_{i-1}(x, y)] \quad (i = 2, 3, \ldots), \tag{26.6}$$

and v_n *is a solution of equation* (23.1) *in case* (b).

PROOF. We first consider case (a). We claim that the equation $x = Tx$ has a unique solution. Indeed, by (25.6),

$$\begin{aligned}\|Tx - Ty\| &= \|RTx + Q_RTx - RTy - Q_RTy\| \\ &= \max\{\|RTx - RTy + P_NQ_RTx - P_NQ_RTy\|, \|Q_{P_N}Tx - Q_{P_N}Ty\| \\ &\leqslant \max\{r\|x-y\| + p\|x-y\|, q\|x-y\|\} = \max\{r+p, q\}\|x-y\|,\end{aligned}$$

and so by (26.1) the operator T is contractive.

Next, since the operator F_k on the right of (23.1) satisfies a Lipschitz condition in x, with constant

$$\Delta_k = r\sum_{i=0}^{k-1} p^i < \frac{r}{1-p} < 1,$$

it follows via the contraction mapping principle that (23.1) is uniquely solvable in E for x_n.

We now prove that the sequence of solutions x_n of (23.1) converges to a solution of the equation

$$x = F_k(x, x). \tag{26.7}$$

In our case, $F_k(x, y) = R_k(RTx, Q_RTy)$, where

$$\begin{aligned}R_1(RTx, Q_RTy) &= RTx + Q_RTy, \\ R_i(RTx, Q_RTy) &= RTx + Q_RTR_{i-1}(RTx, Q_RTy),\end{aligned} \tag{26.8}$$

and so we deduce from (23.1), using (25.6) and (26.1), that

$$\begin{aligned}\|\delta_n\| &= \|R_k(RTx_n, Q_RTx_{n-1}) - R_k(RTx_{n-1}, Q_RTx_{n-2})\| \\ &= \max\{\|RTx_n - RTx_{n-1} + P_NQ_RTF_{k-1}(x_n, x_{n-1}) \\ &\qquad\qquad - P_NQ_RTF_{k-1}(x_{n-1}, x_{n-2})\|, \\ &\|Q_{P_N}TF_{k-1}(x_n, x_{n-1}) - Q_{P_N}TF_{k-1}(x_{n-1}, x_{n-2})\|\} \\ &\leqslant \max\{r\|\delta_n\| + p\|F_{k-1}(x_n, x_{n-1}) - F_{k-1}(x_{n-1}, x_{n-2})\|, \\ &\qquad q\|F_{k-1}(x_n, x_{n-1}) - F_{k-1}(x_{n-1}, x_{n-2})\|\} \quad (\delta_n = x_n - x_{n-1}),\end{aligned}$$

whence

$$\|\delta_n\| \leqslant \frac{p}{1-r}\|F_{k-1}(x_n, x_{n-1}) - F_{k-1}(x_{n-1}, x_{n-2})\|$$

or

$$\|\delta_n\| \leqslant q\|F_{k-1}(x_n, x_{n-1}) - F_{k-1}(x_{n-1}, x_{n-2})\|,$$

i.e.,

$$\begin{aligned}\|\delta_n\| &\leqslant \max\left\{\frac{p}{1-r}, q\right\}\|F_{k-1}(x_n, x_{n-1}) - F_{k-1}(x_{n-1}, x_{n-2})\| \\ &= \varepsilon_1\|F_{k-1}(x_n, x_{n-1}) - F_{k-1}(x_{n-1}, x_{n-2})\|.\end{aligned}$$

Estimating the right-hand side of this inequality in the same way, we get

$$\|\delta_n\| \leqslant \varepsilon_1 \| R_{k-1}(RTx_n - Q_RTx_{n-1}) - R_{k-1}(RTx_{n-1}, Q_RTx_{n-2}) \|$$
$$\leqslant \varepsilon_1 \max \{ r\|\delta_n\| + p\| F_{k-2}(x_n, x_{n-1}) - F_{k-2}(x_{n-1}, x_{n-2}) \|,$$
$$q\| F_{k-2}(x_n, x_{n-1}) - F_{k-2}(x_{n-1}, x_{n-2}) \| \},$$
$$\|\delta_n\| \leqslant \max \left\{ \frac{p\varepsilon_1}{1 - r\varepsilon_1}, q\varepsilon_1 \right\} \| F_{k-2}(x_n, x_{n-1}) - F_{k-2}(x_{n-1}, x_{n-2}) \|$$
$$= \varepsilon_2 \| F_{k-2}(x_n, x_{n-1}) - F_{k-2}(x_{n-1}, x_{n-2}) \|.$$

Repeating the procedure for the right-hand side of this inequality, we finally obtain

$$\|\delta_n\| \leqslant \varepsilon_k \|\delta_{n-1}\|. \tag{26.9}$$

Hence, since $\epsilon_k < 1$ by (26.1), it follows that $\{x_n\}$ is a Cauchy sequence and so has a limit x.

We now show that x is a solution of (26.7). In fact

$$\|x - F_k(x, x)\| \leqslant \|x - x_n\| + \|F_k(x_n, x_{n-1}) - F_k(x, x)\|,$$

so that, in view of the expression for $F_k(x, y)$ and the assumptions of the theorem, we have

$$\|x - F_k(x, x)\| \leqslant \|x - x_n\| + C_k\|x_n - x\| + C_k'\|x_{n-1} - x\|,$$

where C_k and C_k' are constants that depend on k, p and q. Since $\|x_n - x\| \longrightarrow 0$ and $\|x_{n-1} - x\| \longrightarrow 0$, it follows from the last inequality that $\|x - F_k(x, x)\| = 0$, and x is indeed a solution of (26.7).

It is readily seen that x is the unique solution of (26.7). Indeed if there were another solution y, we could prove similarly that $\|x - y\| \leqslant \epsilon_k\|x - y\|$, and hence, since $\epsilon_k < 1$, it would follow that $\|x - y\| = 0$ or $x = y$.

Since each solution of the equation $x = Tx$ is also a solution of (26.7) and we have just proved that each of these equations is uniquely solvable, the convergence of the sequence $\{x_n\}$ to a solution of (26.7) means that the same sequence converges to the solution of the equation $x = Tx$.

The error estimate (26.2) is proved in the usual manner (see §6).

Note that if at least one of the constants p or q is not zero, then $\epsilon_k < \epsilon_l$ for $k > l$, since $\max\{p\epsilon/(1 - r\epsilon), q\epsilon\}$ is then an increasing function on $[0, 1]$. In addition, $\lim \epsilon_k = 0$. Indeed, since ϵ_k is a nonincreasing sequence of nonnegative terms, it has some limit $\epsilon \geqslant 0$. But in view of (26.1) the equality

$$\varepsilon = \max \left\{ \frac{p\varepsilon}{1 - r\varepsilon}, q\varepsilon \right\}$$

can hold only if $\epsilon = 0$. Thus $\lim_{k\to\infty} \epsilon_k = 0$.

In case (b), $x_n = Tz_n$, where z_n is defined by (26.5), and if $v_0 = Tz_0$ then z_n is determined by the same equations as x_n in case (a) (see §9). Thus investigation of algorithm (23.1) in case (b) reduces to investigation of the same algorithm in case (a). The error estimate (26.3) is also established in this way; we need only observe that $\|\delta_n\| \leqslant q_T \|z_n - z_{n-1}\|$.

We now turn to the case that the assumptions of Theorem 26.1 are fulfilled only in a subset D of E and not in the whole space.

Let $Tx = \{T_i x\}$, and suppose that the following inequalities hold in a set D defined by inequalities $\|x_i - g_i\|_{B_i} \leqslant r_i$, where $g = \{g_i\}$ is a fixed element of E:

$$\| P_i T_i x - P_i g_i \|_{B_i} \leqslant F^{(i)}_{RT}(r_1, \ldots, r_m) \qquad (i = 1, 2, \ldots, N),$$
$$\| Q_i T_i x - Q_i g_i \|_{B_i} \leqslant F^{(i)}_{P_N Q_R T}(r_1, \ldots, r_m) \qquad (i = 1, 2, \ldots, N),$$
$$\| T_i x - g_i \|_{B_i} \leqslant F^{(i)}_{Q P_N T}(r_1, \ldots, r_m) \qquad (i = N+1, \ldots, m).$$

If there exist numbers $r_i = r_i^* > 0$ such that

$$F^{(i)}_{RT}(r_1, \ldots, r_m) + F^{(i)}_{P_N Q_R T}(r_1, \ldots, r_m) \leqslant r_i \qquad (i = 1, \ldots, N),$$
$$F^{(i)}_{Q P_N T}(r_1, \ldots, r_m) \leqslant r_i \qquad (i = N+1, \ldots, m), \tag{26.10}$$

and the assumptions of Theorem 26.1 hold on the set $D^*(\|x_i - g_i\|_{B_i} \leqslant r_i^*)$, then the conclusion of the theorem for algorithm (23.1) in case (a) is also valid, provided that $x_0 \in D^*$. Moreover, if we define x_0 in case (b) by $x_0 = Tz_0 (z_0 \in D^*)$, then the theorem is also true in D^* for case (b).

Now consider case (a), and suppose that x_0 is specified independently of z_0. If there exist numbers $r_i = r_i^{**} > 0$ satisfying (26.10) and the system of inequalities

$$F^{(i)}_{RT}(r_1, \ldots, r_m) + \| Q_i (x_{i,0} - g_i) \| \leqslant r_i \ (i = 1, \ldots, N);$$
$$\| x_{i,0} - g_i \|_{B_i} \leqslant r_i \quad (i = N+1, \ldots, m), \tag{26.11}$$

and the assumptions of Theorem 26.1 hold in the ball D^{**} $(\|x_i - g_i\| \leqslant r^{**})$, then the theorem is also true for algorithm (23.1) in case (b) in the set D^*.

We shall not present a detailed proof of these statements. We only observe that under these conditions, for any $x, y \in D^*$

$$RTx + Q_R Ty \in D^*,$$

and for $x, y \in D^{**}$

$$RTx + Q_R x_0 \in D^{**}, \qquad RTx + Q_R Ty \in D^{**}.$$

But this means (see §9) that the operators on the right of (23.1) (case (a)) and on the right of the equations for z_n (case (b)) map the appropriate regions into themselves, and all the arguments figuring in the proof of Theorem 26.1 may be applied, with the sole difference that the space E is replaced by its subsets D^* and D^{**}.

THEOREM 26.2. *Let the B_i be Hilbert spaces and P_i orthogonal projections. If*

$$\max\{r^2 + p^2, q^2\} < 1, \tag{26.12}$$

then the successive approximations x_n are uniquely determined by equations (23.1) *in both cases* (a) *and* (b), *and they converge to the unique solution in E of the equation $x = Tx$, with error estimates* (26.2) *and* (26.3), *respectively; the quantities ϵ_k may be defined as*

$$\varepsilon_i = \max\left\{\frac{p\varepsilon_{i-1}}{\sqrt{1 - r^2\varepsilon_{i-1}^2}},\ q\varepsilon_{i-1}\right\} (\varepsilon_0 = 1). \tag{26.13}$$

As in the general case, $\epsilon_k < \epsilon_l$ for $k > l$, provided at least one of the constants p or q is not zero.

The proof is entirely analogous to that of Theorem 26.1, except that in estimating $\|\delta_n\|^2$ one uses (25.7) and (26.12) instead of (25.6) and (26.1), respectively.

If the assumptions of Theorem 26.2 hold only in some subset of E, the same conclusions hold as in the general case, with (26.10) and (26.11) replaced respectively by

$$\begin{gathered}[F_{RT}^{(i)}(r_1, \ldots, r_m)]^2 + [F_{P_NQ_RT}^{(i)}(r_1, \ldots, r_m)]^2 \leqslant r_i^2 \\ (i = 1, \ldots, N), \\ F_{QP_NT}^{(i)}(r_1, \ldots, r_m) \leqslant r_i \qquad (i = N+1, \ldots, m)\end{gathered} \tag{26.14}$$

and

$$\begin{gathered}[F_{RT}^{(i)}(r_1, \ldots, r_m)]^2 + \|Q_i(x_{i,0} - g_i)\|_{B_i}^2 \leqslant r_i^2 \qquad (i = 1, \ldots, N), \\ \|x_{i,0} - g_i\|_{B_i} \leqslant r_i \qquad (i = N+1, \ldots, m).\end{gathered} \tag{26.15}$$

§27. Case of k-normed direct-product Banach space with projection R. Error estimate for projection method

Let E be a space whose elements are sequences $x = \{x_i\}$ $(i = 1, \ldots, m)$, where the x_i are elements of Banach spaces B_i, respectively. Define a norm in E by

$$\|x\| = \sqrt{\sum_{i=1}^{m} \|x_i\|_{B_i}^2}\,. \tag{27.1}$$

As usual $\|x_i\|_{B_i}$ denotes the norm in the appropriate Banach space B_i.

Consider the projections P_N, Π and R, defined as in previous sections.

For any $x, y \in E$, we have

$$\|Rx + Q_Ry\|^2 \leqslant \|Rx\|^2 + Q_Ry\|^2 + 2\|Rx\|\|P_NQ_Ry\|. \tag{27.2}$$

In fact, by the definitions of R and $Q_R = I - R$, we have

$$Rx + Q_R y\|^2 = \sum_{i=1}^{N} \|P_i x_i + Q_i y_i\|_{B_i}^2 + \sum_{i=N+1}^{m} \|y_i\|_{B_i}^2$$

$$\leqslant \sum_{i=1}^{N} (\|P_i x_i\|_{B_i}^2 + 2\|P_i x_i\|_{B_i}\|Q_i y_i\|_{B_i} + \|Q_i y_i\|_{B_i}^2) + \sum_{=N+1}^{m} \|y_i\|_{B_i}^2$$

$$\leqslant \sum_{i=1}^{N} \|P_i x_i\|_{B_i}^2 + 2\sqrt{\sum_{i=1}^{N} \|P_i x_i\|_{B_i}^2}\sqrt{\sum_{i=1}^{N} \|Q_i y_i\|_{B_i}^2}.$$

$$+ \sum_{i=1}^{N} \|Q_i y_i\|_{B_i}^2 + \sum_{i=N+1}^{m} \|y_i\|_{B_i}^2,$$

whence we at once obtain (27.2).

If the B_i are Hilbert spaces, then E is also a Hilbert space, with scalar product

$$(x, y) = \sum_{i=1}^{m} (x_i, y_i)_{B_i} \tag{27.3}$$

where $(x_i, y_i)_{B_i}$ denotes the scalar product in B_i.

In this case, R is a selfadjoint operator if all the P_i's are selfadjoint. Indeed,

$$(Rx, y) = \sum_{i=1}^{N} (P_i x_i, y_i)_{B_i} = \sum_{i=1}^{N} (x_i, P_i y_i)_{B_i} = (x, Ry). \tag{27.4}$$

But if R is selfadjoint, we have a valid analog of the equality considered in §16:

$$\|Rx + Q_R y\|^2 = \|Rx\|^2 + \|Q_R y\|^2. \tag{27.5}$$

Thus R is a special case of the orthogonal projections considered in §16.

Henceforth we shall assume only that the B_i's are Banach spaces. We shall develop an error estimate for the projection method applied to the equation $x = Tx$, with a projection R in a k-normed space; in other words, we wish to estimate $\|x - y\|$, where x is an exact solution of the equation $x = Tx$ and y a solution of the equation

$$y = RTy. \tag{27.6}$$

THEOREM 27.1. *Let D be a subset of E in which the equations $x = Tx$ and (27.6) have solutions x and y, respectively, and the operators RT, $Q_R T$ and $P_N Q_R T$ satisfy Lipschitz conditions with Lipschitz constants q_{RT}, $q_{Q_R T}$ and $q_{P_N Q_R T}$ such that*

$$q_{RT}^2 + q_{Q_R T}^2 + 2q_{P_N Q_R T}^2 < 1. \tag{27.7}$$

Then an estimate for the error $\|x - y\|$ is given by

$$\|x - y\| \leqslant \frac{a}{p} + \frac{1}{p}\sqrt{a^2 + p\|Q_R T y\|^2}, \tag{27.8}$$

where $a = q_{Q_R T}\|Q_R Ty\| + q_{RT}\|P_N Q_R Ty\|$ and $p = 1 - q_{RT}^2 - q_{Q_R T}^2 - 2q_{P_N Q_R T} q_{RT}$.

PROOF. Since x and y are solutions of the respective equations $x = Tx$ and (27.6),

$$x - y = Tx - RTy + RTx - RTx = Q_R Tx + RTx - RTy,$$

and so, by (27.2), we have

$$\begin{aligned}\|x-y\|^2 &\leqslant \|Q_R Tx\|^2 + \|RTx - RTy\|^2 + 2\|RTx - RTy\|\|P_N Q_R Tx\| \\ &\leqslant \|Q_R Tx\|^2 + q_{RT}^2\|x-y\|^2 + 2\|P_N Q_R Tx\| q_{RT}\|x-y\|. \end{aligned} \tag{27.9}$$

Applying the inequalities

$$\|Q_R Tx\| \leqslant \|Q_R Tx - Q_R Ty\| + \|Q_R Ty\| \leqslant q_{Q_R T}\|x-y\| + \|Q_R Ty\|,$$
$$\|P_N Q_R Tx\| \leqslant q_{P_N Q_R T}\|x-y\| + \|P_N Q_R Ty\|$$

and substituting them in the right-hand side of (27.9), we obtain

$$\begin{aligned}\|x-y\|^2 \leqslant{}& (q_{RT}^2 + q_{Q_R T}^2 + 2q_{P_N Q_R T} q_{RT})\|x-y\|^2 \\ &+ 2(q_{Q_R T}\|Q_R Ty\| + q_{RT}\|P_N Q_R Ty\|)\|x-y\| + \|Q_R Ty\|^2.\end{aligned}$$

Solving this inequality for $\|x - y\|$, we get (27.8).

§28. Investigation of generalized iterative projection methods for operator equations in a k-normed space with projection operator R

We now present some convergence criteria and error estimates for generalized iterative projection methods applied to the equation $x = Tx$ in a k-normed direct product Banach space E (see (27.1)), with the projection R of §25.

We shall consider successive approximations defined by (23.1) in two cases:

a) $F(x, y) = RTx + Q_R Ty$,

b) $F(x, y) = T(Rx + Q_R y)$.

Using property (27.2), which is valid in this case, we prove

THEOREM 28.1. *If*

$$r^2 + 2ps + s^2 < 1, \tag{28.1}$$

where $r = q_{RT}$, $p = q_{P_N Q_R T}$ *and* $s = q_{Q_R T}$ *(as usual, the q's denote Lipschitz constants for the appropriate operators), then equations* (23.1) *are uniquely solvable for* x_n *in both cases* (a) *and* (b), *and the sequence* $\{x_n\}$ *converges to the unique solution* x *of the equation* $x = Tx$ *in* E, *with error estimates* (26.2) *and* (26.3), *respectively, with* z_n *defined as in* §26 *and* ϵ_i *defined recursively by*

$$\varepsilon_i = \frac{rp\varepsilon_{i-1} + \sqrt{r^2p^2\varepsilon_{i-1}^2 + (1 - r^2\varepsilon_{i-1}^2)s^2}}{1 - r^2\varepsilon_{i-1}^2}\varepsilon_{i-1} \quad (\varepsilon_0 = 1). \tag{28.2}$$

Moreover, $\epsilon_k < \epsilon_l$ *for* $k > l$ *and* $\epsilon_l \neq 0$, $\epsilon_k \to 0$ *as* $k \to \infty$.

PROOF. We first show that under condition (28.1) the equation $x = Tx$ is uniquely solvable. We have

$$\|Tx - Ty\| = \|RTx + Q_RTx - RTy - Q_RTy\|$$
$$\leqslant \|RTx - RTy\|^2 + 2\|RTx - RTy\|\|R_NQ_RTx - P_NQ_RTy\|$$
$$+ \|Q_RTx - Q_RTy\|^2 \leqslant (r^2 + 2rp + s^2)\|x - y\|,$$

and so, by (28.1), we can apply the contraction mapping principle.

Now it is readily seen that for any $x, y, z \in E$

$$\|R_k(RTx, Q_RTz) - R_k(RTy, Q_RTz)\| \leqslant \Delta_k\|x - y\|,$$

where the constants Δ_i are defined recursively by

$$\Delta_i = r^2 + 2rp\sqrt{\Delta_{i-1}} + s^2\Delta_{i-1}\,(\Delta_1 = r^2), \tag{28.3}$$

$\Delta_k < 1$, and R_k is defined as in §26. Consequently, in case (a) the contraction mapping principle is valid for equations (23.1) and so they are uniquely solvable in E.

We now prove that the solutions of equations (23.1) in case (a) converge to the solution of $x = Tx$. Case (b) may be reduced to case (a) (see §26).

Noting that

$$F_k(x, y) = R_k(RTx, Q_RTy) = RTx + Q_RTF_{k-1}(x, y),$$

and using (27.2), we deduce from (23.1) that

$$\|\delta_n\|^2 \leqslant \|RTx_n - RTx_{n-}\|^2 + \|Q_RTF_{k-1}(x_n, x_{n-1}) - Q_RTF_{k-1}(x_{n-1}, x_{n-2})\|^2$$
$$+ 2\|RTx_n - RTx_{n-1}\| \times \|P_NQ_RTF_{k-1}(x_n, x_{n-1}) - P_NQ_RTF_{k-1}(x_{n-1}, x_{n-2})\|$$
$$\leqslant r^2\|\delta_n\|^2 + s^2\|F_{k-1}(x_n, x_{n-1}) - F_{k-1}(x_{n-1}, x_{n-2})\|^2$$
$$+ 2rp\|\delta_n\|\|F_{k-1}(x_n, x_{n-1}) - F_{k-1}(x_{n-1}, x_{n-2})\|.$$

Solving for $\|\delta_n\|$, we find that

$$\|\delta_n\| \leqslant \frac{rp + \sqrt{r^2p^2 + (1 - r^2)s^2}}{1 - r^2}\|F_{k-1}(x_n, x_{n-1}) - F_{k-1}(x_{n-1}, x_{n-2})\|$$
$$= \varepsilon_1\|F_{k-1}(x_n, x_{n-1}) - F_{k-1}(x_{n-1}, x_{n-2})\|.$$

We now estimate $\|F_{k-1}(x_n, x_{n-1}) - F_{k-1}(x_{n-1}, x_{n-2})\|$ in the same way as before, to obtain

$$\|\delta_n\|^2 \leqslant \varepsilon_1^2[\|RTx_n - RTx_{n-1}\|^2 + \|Q_RTF_{k-2}(x_n, x_{n-1})$$
$$- Q_RTF_{k-2}(x_{n-1}, x_{n-2})\|^2 + 2\|RTx_n - RTx_{n-1}\|$$
$$\times \|P_NQ_RTF_{k-2}(x_n, x_{n-1}) - P_NQ_RTF_{k-2}(x_{n-1}, x_{n-2})\|]$$
$$\leqslant \varepsilon_1^2[r^2\|\delta_n\|^2 + s^2\|F_{n-2}(x_n, x_{n-1}) - F_{k-2}(x_{n-1}, x_{n-2})\|^2$$
$$+ 2rp\|\delta_n\|\|F_{k-2}(x_n, x_{n-1}) - F_{k-2}(x_{n-1}, x_{n-2})\|],$$

whence

$$\|\delta_n\| \leqslant \frac{rp\varepsilon_1^2 + \sqrt{r^2p^2\varepsilon_1^4 + (1 - r^2\varepsilon_1^2)s^2\varepsilon_1^2}}{1 - r^2\varepsilon_1^2}\|F_{k-2}(x_n, x_{n-1}) - F_{k-2}(x_{n-1}, x_{n-2})\|$$
$$= \varepsilon_2\|F_{k-2}(x_n, x_{n-1}) - F_{k-2}(x_{n-1}, x_{n-2})\|.$$

Repeating the argument for the right-hand side of this inequality, we finally obtain

$$\| \delta_n \| \leqslant \varepsilon_k \| \delta_{n-1} \|, \tag{28.4}$$

where ϵ_k is defined by (28.2).

It is readily seen that

$$\psi(\varepsilon) = \frac{rp\varepsilon + \sqrt{r^2p^2\varepsilon^2 + (1 - r^2\varepsilon^2)s^2}}{1 - r^2\varepsilon^2},$$

is an increasing function on $[0, 1]$, provided it does not vanish there identically, and its maximum value is ϵ_1, which by (28.1) is less than 1. Consequently, $\epsilon_k < \epsilon_l$ for $k > l$, provided that $\epsilon_l \neq 0$. It is also clear that $\epsilon_k \leqslant \epsilon_1^k$, and if $\epsilon_1 \neq 0$ then $\epsilon_k < \epsilon_1^k$.

Thus, by (28.4) and the inequality $\epsilon_k < 1$, the sequence $\{x_n\}$ of solutions of (23.1) (case (a)) is a Cauchy sequence and since the space E is complete it converges to a unique limit $x \in E$.

We claim that x is the unique solution of the equation $x = F_k(x, x)$. It is readily seen, estimating the quantity $\|F_k(x_n, x_{n-1}) - F_k(x, x)\|$, that

$$\| x - F_k(x, x) \| \leqslant \| x - x_n \| + \| F_k(x_n, x_{n-1}) - F_k(x, x) \|$$
$$\leqslant \| x - x_n \| + \sqrt{r^2 + 2rp + s^2} \max \{\| x_n - x \|, \| x_{n-1} - x \|\},$$

and so as $\|x_n - x\| \to 0$ we have $\|x - F_k(x, x)\| = 0$, so that x is indeed a solution of the equation $x = F_k(x, x)$. For any other solution y of this equation, we prove in the same way as (28.4) that $\|x - y\| \leqslant \epsilon_k \|x - y\|$, whence, since $\epsilon_k < 1$, it follows that $\|x - y\| = 0$ and $x = y$.

Since each solution of the equation $x = Tx$ is a solution of the equation $x = F_k(x, x)$, it follows that $\{x_n\}$ converges to the solution of $x = Tx$.

The error estimates are deduced from (28.4) by the same reasoning as in the general case (see §6).

Finally, we consider the case that the assumptions of Theorem 28.1 hold not throughout the space E but only in some ball.

Suppose that for all $x \in S(g, r)$

$$\| RTx - Rg \| \leqslant F_{RT}(r), \qquad \| Q_RTx - Q_Rg \| \leqslant F_{Q_RT}(r),$$
$$\| P_NQ_RTx - P_NQ_Rg \| \leqslant F_{P_NQ_RT}(r),$$

where the F's are certain functions.

If there exists $r = r^* > 0$ such that

$$F_{RT}^2(r) + 2F_{RT}(r) F_{P_NQ_RT}(r) + F_{Q_RT}^2(r) \leqslant r^2, \tag{28.5}$$

and the assumptions of Theorem 28.1 hold in $S(g, r^*)$, then (23.1) and the equation for z_n in case (b) (see §26) are uniquely solvable in $S(g, r^*)$ for x_n, provided that $x_0, z_0 \in S(g, r^*)$, and the other conclusions of the theorem remain valid in $S(g, r^*)$.

However, if the initial approximation in case (b) is specified independently of z_0, then the radius of a ball with center g in which the equations for z_n are uniquely solvable may be defined as a positive number $r = r^{**}$ such that

$$F_{RT}^2(r) + 2F_{RT}(r) F_{P_NQ_RT}(r) + F_{Q_RT}^2(r) \leqslant r^2,$$
$$F_{RT}^2(r) + 2F_{RT}(r) \| P_N Q_R (x_0 - g) \| + \| Q_R (x_0 - g) \|^2 \leqslant r^2. \quad (28.6)$$

In this situation, if the assumptions of Theorem 28.1 hold in $S(g, r^{**})$, then the conclusion of the theorem with regard to convergence of the algorithm (23.1) (case (b)) remains valid for $S(g, r^{**})$.

§29. Case of l-normed direct-product Banach space with projection P_N. Investigation of iterative projection methods

As in §§22–28, the elements of the space E are sequences $x = \{x_i\}$ ($i = 1, \ldots, m$), where the x_i are elements of Banach spaces B_i, with addition and multiplication by a scalar defined as in §22.

In the case $m = \infty$, we shall assume that the series $\Sigma_1^\infty \|x_i\|_{B_i}$ are all convergent (where $\|x_i\|_{B_i}$, as usual, denotes the norm of x_i in B_i). We then define the norm of an element $x \in E$ by

$$\|x\| = \sum_{i=1}^m \|x_i\|_{B_i}.$$

By analogy with the case of the classical sequence space, we shall call this an l-norm.

The projections P_N and $Q_N = I - P_N$ are defined as in §22.

Analyzing the expression for $P_N x + Q_N y$, we readily see that

$$\| P_N x + Q_N y \| = \| P_N x \| + \| Q_N y \|. \quad (29.1)$$

Note that in the general case of a Banach space the equality in (29.1) must be replaced by an inequality $\leqslant$.

Property (29.1) may also be utilized to derive convergence criteria and error estimates for iterative projection methods which do not follow from the general theory of Chapter I.

We consider the algorithms

a) $u_n = R_k (P_N T u_n, Q_N T u_{n-1}) \, (u_0 \in E, \; n = 1, 2, \ldots)$, (29.2)

b) $v_n = S_k (P_N v_n, Q_N v_{n-1}) \quad (v_0 \in E, \; n = 1, 2, \ldots)$. (29.3)

As before, the operators R_k and S_k are defined by

$$R_1 (P_N T x, Q_N T y) = P_N T x + Q_N T y,$$
$$R_i (P_N T x, Q_N T y) = P_N T x + Q_N T R_{i-1} (P_N T x, Q_N T y) \quad (29.4)$$
$$(i = 2, 3, \ldots),$$

$$S_1(P_Nx, Q_Ny) = T(P_Nx + Q_Ny),$$
$$S_i(P_Nx, Q_Ny) = T[P_Nx + Q_NS_{i-1}(P_Nx, Q_Ny)] \quad (i = 2, 3, \ldots). \tag{29.5}$$

THEOREM 29.1. *If the operators T, P_NT and Q_NT satisfy Lipschitz conditions with constants q_T, q_{P_NT} and q_{Q_NT} such that*

$$q_T \sum_{i=0}^{k-1} q_{Q_NT}^i < 1, \tag{29.6}$$

then equations (29.2) and (29.3) are uniquely solvable in E for u_n and v_n, and the sequences $\{u_n\}$ and $\{v_n\}$ converge to a solution of the equation $x = Tx$, with error estimates

$$\|x - u_n\| \leqslant N_k(1-\varepsilon_k)^{-1}\varepsilon_k^{n-p}\|Q_N(Tu_p - Tu_{p-1})\| \quad (1 \leqslant p \leqslant n), \tag{29.7}$$
$$\|x - v_n\| \leqslant q_T N_k(1-\varepsilon_k)^{-1}\varepsilon_k^{n-p}\|Q_N(v_p - v_{p-1})\| \quad (1 \leqslant p \leqslant n), \tag{29.8}$$

where

$$N_k = q_{Q_NT}^{k-1}\left(1 - q_{P_NT}\sum_{i=0}^{k-1} q_{Q_NT}^i\right)^{-1}, \qquad \varepsilon_k = q_T q_{Q_NT}^{k-1}.$$

PROOF. It follows from (29.1) that $q_{P_NT} \leqslant q_T$ and $q_{Q_NT} \leqslant q_T$. Indeed, we have

$$\|P_NTx - P_NTy\| + \|Q_NTx - Q_NTy\| = \|Tx - Ty\|,$$

whence

$$\|P_NTx - P_NTy\| \leqslant \|Tx - Ty\| \leqslant q_T\|x - y\|,$$
$$\|Q_NTx - Q_NTy\| \leqslant \|Tx - Ty\| \leqslant q_T\|x - y\|.$$

Next, it is readily seen that the operators on the right-hand sides of equations (29.2) and the equation

$$P_Nv_n = P_NS_k(P_Nv_n, Q_Nv_{n-1}), \tag{29.9}$$

obtained from (29.3) by applying P_N to both sides, satisfy Lipschitz conditions with constants bounded above by

$$\Delta_k = q_{P_NT}\sum_{i=0}^{k-1} q_{Q_NT}^i.$$

But it follows from the inequality $q_{P_NT} \leqslant q_T$ and from (29.6) that $\Delta_k < 1$, and so we may apply the contraction mapping principle to (29.2) and (29.9). Consequently, each of equations (29.2) and (29.3) is uniquely solvable for u_n and v_n.

We now show that the sequences $\{u_n\}$ and $\{v_n\}$ converge to a solution of the equation $x = Tx$. Setting $\delta_{z_n} = z_n - z_{n-1}$, we deduce from (29.2) and (29.3) that

$$\|\delta Tu_n\| \leqslant q_T q_{Q_NT}^{k-1}\|Q_N\delta Tu_{n-1}\| + q_T\sum_{i=0}^{k-1} q_{Q_NT}^i\|P_N\delta Tu_n\|, \tag{29.10}$$

$$\|\delta v_n\| \leqslant q_T q_{Q_NT}^{k-1} \|Q_N \delta v_{n-1}\| + q_T \sum_{i=0}^{k-1} q_{Q_NT}^i \|P_N \delta v_n\|. \tag{29.11}$$

By (29.1), we have

$$\|\delta T u_n\| = \|Q_N \delta T u_n\| + \|P_N \delta T u_n\|, \qquad \|\delta v_n\| = \|Q_N \delta v_n\| + \|P_N \delta v_n\|.$$

Now, substituting these expressions for $\|\delta T u_n\|$ and $\|\delta v_n\|$ into the left-hand sides of (29.10) and (29.11), respectively, and using (29.6), we can sharpen the resulting inequalities. The result is

$$\|Q_N \delta T u_n\| \leqslant q_T q_{Q_NT}^{k-1} \|Q_N \delta T u_{n-1}\|, \tag{29.12}$$

$$\|Q_N \delta v_n\| \leqslant q_T q_{Q_NT}^{k-1} \|Q_N \delta v_{n-1}\|. \tag{29.13}$$

It follows from these estimates that $\{Q_N T u_n\}$ and $\{Q_N v_n\}$ are Cauchy sequences. It is readily seen that the same holds for $\{u_n\}$ and $\{v_n\}$, and that these solutions converge to a solution of the equations $x = Tx$.

To derive the error estimates (29.7) and (29.8) it suffices to observe that under the assumptions of the theorem

$$\|\delta u_n\| \leqslant N_k \|Q_N \delta T u_{n-1}\|, \qquad \|\delta v_n\| \leqslant q_T N_k \|Q_N \delta v_{n-1}\|.$$

CHAPTER III

APPLICATION OF ITERATIVE PROJECTION METHODS TO SOME SPECIAL CLASSES OF OPERATOR EQUATIONS

§30. Finite and infinite systems of algebraic and transcendental equations

Consider a system of algebraic (or transcendental) equations

$$\begin{aligned} x_1 &= \varphi_1(x_1, x_2, \ldots, x_m), \\ x_2 &= \varphi_2(x_1, x_2, \ldots, x_m), \\ &\cdots\cdots\cdots\cdots \\ x_m &= \varphi_m(x_1, x_2, \ldots, x_m) \end{aligned} \tag{30.1}$$

or, in abbreviated notation,

$$x_i = \varphi_i(x_1, x_2, \ldots, x_m) \qquad (i = 1, 2, \ldots, m) \tag{30.2}$$

where the φ_i are real functions.

The set of arguments $x_1, \ldots, x_m$ may be viewed as an m-vector in a suitable vector space.

The set of functions $\{\varphi_i\}$ is also an m-vector, or vector-valued function, depending on the previous vector. We denote it by Φ: $\Phi = \{\varphi_1, \ldots, \varphi_m\}$, and thus system (30.1) (or (30.2)) may be expressed as an operator equation:

$$x = \Phi x. \tag{30.3}$$

In other words, this is a nonlinear operator equation of the type considered in the previous chapters.

We shall assume the following conditions:

a) The functions φ_i are jointly continuous in the domain $-\infty < x_i < \infty$.

b) The functions φ_i have continuous and bounded partial derivatives with respect to each of the variables in the same domain:

$$|\varphi'_{ix_j}(x_1, x_2, \ldots, x_m)| \leqslant m_{ij} \leqslant M < \infty \quad (i, j = 1, 2, \ldots, m). \tag{30.4}$$

Under these assumptions, we may treat system (30.1) as a nonlinear operator equation (30.3) in the space of m-bounded number sequences, with the norm $\|x\|$ of an element $x = (x_1, \ldots, x_m)$ defined by

$$\|x\| = \max_i |x_i|. \tag{30.5}$$

This is a special case of the more general m-normed direct-product Banach space considered in Chapter II.

Let us treat system (30.1) by the projection and iterative projection methods, letting the projection P be the operator P_N of §§22–24.

Since $P_N x = \{x_i^{(P_N)}\}$, where

$$x_i^{(P_N)} = \begin{cases} x_i & \text{for } i = 1, 2, \ldots, N, \\ 0 & \text{for } i = N+1, \ldots, m, \end{cases}$$

it follows that in our case the equation (22.9) of the projection method is a system of equations

$$\begin{aligned} x_i &= \varphi_i(x_1, x_2, \ldots, x_m) \quad (i = 1, 2, \ldots, N), \\ x_i &= 0 \quad (i = N+1, \ldots, m). \end{aligned} \tag{30.6}$$

The equation (9.10) of the iterative projection method for $k = 1$ may be written as a system

$$\begin{aligned} x_{i,n} &= \varphi_i(x_{1,n}, x_{2,n}, \ldots, x_{m,n}) \quad (i = 1, 2, \ldots, N), \\ x_{i,n} &= \varphi_i(x_{1,n}, \ldots, x_{N,n}, x_{N+1,n-1}, \ldots, x_{m,n-1}) \\ &\qquad (i = N+1, \ldots, m). \end{aligned} \tag{30.7}$$

If we use the modified iterative projection method defined by a system of operator equations (12.2), the system for the vectors $u_n = \{u_{i,n}\}$, $v_n = \{v_{i,n}\}$ $(k = 1)$ is

$$\begin{aligned} u_{i,n} &= \varphi_i(u_{1,n} + v_{1,n}, \ldots, u_{m,n} + v_{m,n}) \quad (i = 1, 2, \ldots, N), \\ u_{i,n} &= 0 \qquad (i = N+1, \ldots, m), \\ v_{i,n} &= 0 \qquad (i = 1, 2, \ldots, N), \\ v_{i,n} &= \varphi_i(u_{1,n} + v_{1,n-1}, \ldots, u_{m,n} + v_{m,n-1}) \quad (i = N+1, \ldots, m). \end{aligned} \tag{30.8}$$

The condition $\Psi(Px + Qy) = \Psi Px + \Psi Qy$ will hold, for example, if $\Psi = \{\psi_i\}$ $(i = 1, \ldots, m)$,

$$\psi_i(x_1, x_2, \ldots, x_m) = \sum_{j=1}^{m} f_{i,j}(x_j), \quad f_{i,j}(0) = 0,$$

where the $f_{i,j}$ are certain functions. If $\Phi x = a + \Psi x$, we can use algorithm (12.24). The corresponding system of equations for the vectors $u_n = \{u_{i,n}\}$ and $v_n = \{v_{i,n}\}$ is

$$u_{i,n} = a_i + \sum_{j=1}^{m} f_{i,j}(u_{j,n} + v_{j,n}) \qquad (i = 1, 2, \ldots, N),$$

$$u_{i,n} = a_i + \sum_{j=1}^{N} f_{i,j}(u_{j,n} + v_{j,n}) \quad (i = N+1, \ldots, m), \tag{30.9}$$

$$v_{i,n} = 0 \quad (i = 1, 2, \ldots, N),$$

$$v_{i,n} = \sum_{j=N+1}^{m} f_{i,j}(u_{j,n} + v_{j,n-1}) \quad (i = N+1, \ldots, m).$$

Let us evaluate Lipschitz constants for the various operators involved. Since $Tx = \{\varphi_i x\}$, we see, setting $c_{ij} = \sup|\partial\varphi_i/\partial x_j|$, that

$$\|Tx - Ty\| = \max_i |\varphi_i(x_1, x_2, \ldots, x_m) - \varphi_i(y_1, y_2, \ldots, y_m)| \tag{30.10}$$

$$\leqslant \max_i \sum_{j=1}^{m} c_{ij}|x_j - y_j| \leqslant \max_i \sum_{j=1}^{m} c_{ij} \max_j |x_j - y_j| = \max_i \sum_{j=1}^{m} c_{ij}\|x - y\|.$$

Consequently, the Lipschitz constant q_T for the operator T is bounded by

$$\bar{q}_T = \max_i \sum_{j=1}^{m} c_{ij}.$$

Similarly, we find Lipschitz constants for the other operators:

$$q_{PT} \leqslant \bar{q}_{PT} = \max_{i=1,\ldots,N} \sum_{j=1}^{m} c_{ij}, \quad q_{QT} \leqslant \bar{q}_{QT} = \max_{i=N+1,\ldots,m} \sum_{j=1}^{m} c_{ij},$$

$$q_{TP} \leqslant \bar{q}_{TP} = \max_{i=1,\ldots,m} \sum_{j=1}^{N} c_{ij}, \quad q_{TQ} \leqslant \bar{q}_{TQ} = \max_{i=1,\ldots,m} \sum_{j=N+1}^{m} c_{ij},$$

$$q_{PTP} \leqslant \bar{q}_{PTP} = \max_{i=1,\ldots,N} \sum_{j=1}^{N} c_{ij}, \quad q_{QTP} \leqslant \bar{q}_{QTP} = \max_{i=N+1,\ldots,m} \sum_{j=1}^{N} c_{ij},$$

$$q_{PTQ} \leqslant \bar{q}_{PTQ} = \max_{i=1,\ldots,N} \sum_{j=N+1}^{m} c_{ij}, \quad q_{QTQ} \leqslant \bar{q}_{QTQ} = \max_{i=N+1,\ldots,m} \sum_{j=N+1}^{m} c_{ij}.$$

If we treat (30.2) as a nonlinear operator equation of type (30.3) in a vector space normed by

$$\|x\| = \sum_{i=1}^{m} |x_i|, \tag{30.11}$$

we have

$$\|Tx - Ty\| = \sum_{i=1}^{m} |\varphi_i(x_1, x_2, \ldots, x_m) - \varphi_i(y_1, y_2, \ldots, y_m)|$$

$$\leqslant \sum_{i=1}^{m}\sum_{j=1}^{m} c_{ij}|x_j - y_j| \leqslant \max_j \sum_{i=1}^{m} c_{ij} \sum_{j=1}^{m} |x_j - y_j| \tag{30.12}$$

$$= \max_j \sum_{i=1}^{m} c_{ij}\|x - y\|.$$

Consequently, an estimate for the Lipschitz constant of the operator $T = \Phi$ is provided in this space by

$$q_T \leqslant \bar{q}_T = \max_{j=1,\dots,m} \sum_{i=1}^{m} c_{ij}.$$

Proceeding in analogous fashion for the other operators, we obtain

$$q_{PT} \leqslant \bar{q}_{PT} = \max_{j=1,\dots,m} \sum_{i=1}^{N} c_{ij}, \qquad q_{QT} \leqslant \bar{q}_{QT} = \max_{j=1,\dots,m} \sum_{i=N+1}^{m} c_{ij},$$

$$q_{TP} \leqslant \bar{q}_{TP} = \max_{j=1,\dots,N} \sum_{i=1}^{m} c_{ij}, \qquad q_{TQ} \leqslant \bar{q}_{TQ} = \max_{j=N+1,\dots,m} \sum_{i=1}^{m} c_{ij},$$

$$q_{PTP} \leqslant \bar{q}_{PTP} = \max_{j=1,\dots,N} \sum_{i=1}^{N} c_{ij}, \qquad q_{QTP} \leqslant \bar{q}_{QTP} = \max_{j=1,\dots,N} \sum_{i=N+1}^{m} c_{ij},$$

$$q_{PTQ} \leqslant \bar{q}_{PTQ} = \max_{j=N+1,\dots,m} \sum_{i=1}^{N} c_{ij}, \qquad q_{QTQ} \leqslant \bar{q}_{QTQ} = \max_{j=N+1,\dots,m} \sum_{i=N+1}^{m} c_{ij}.$$

We now assume that the functions φ_i in system (30.2) satisfy the condition

$$|\varphi_i(x_1, x_2, \dots, x_m) - \varphi_i(y_1, y_2, \dots, y_m)|^2 \leqslant \sum_{j=1}^{m} c'_{ij} |x_j - y_j|^2, \tag{30.13}$$

and treat system (30.2) as a nonlinear operator equation (30.3) in the space l_2 of sequences normed by

$$\|x\|^2 = \sum_{i=1}^{m} |x_i|^2. \tag{30.14}$$

Estimating the Lipschitz constant of T, we have

$$\begin{aligned} \|Tx - Ty\|^2 &= \sum_{i=1}^{m} |\varphi_i(x_1, x_2, \dots, x_m) - \varphi_i(y_1, y_2, \dots, y_m)|^2 \\ &\leqslant \sum_{i=1}^{m} \sum_{j=1}^{m} c'_{ij} |x_j - y_j|^2 \leqslant \max_j \sum_{i=1}^{m} c'_{ij} \sum_{j=1}^{m} |x_j - y_j|^2 = \max_j \sum_{i=1}^{m} c'_{ij} \|x - y\|^2. \end{aligned} \tag{30.15}$$

Consequently,

$$q_T^2 \leqslant \bar{q}_T^2 = \max_j \sum_{i=1}^{m} c'_{ij}.$$

If the estimates for the Lipschitz constants are expressed in terms of bounds for the derivatives $|\partial\varphi_i/\partial x_j|$, use of Cauchy's inequality gives

$$\begin{aligned} \|Tx - Ty\|^2 &= \sum_{i=1}^{m} |\varphi_i(x_1, x_2, \dots, x_m) - \varphi_i(y_1, y_2, \dots, y_m)|^2 \\ &\leqslant \sum_{i=1}^{m} \left(\sum_{j=1}^{m} c_{ij} |x_j - y_j| \right)^2 \leqslant \sum_{i=1}^{m} \left(\sum_{j=1}^{m} c_{ij}^2 \sum_{j=1}^{m} |x_j - y_j|^2 \right) = \sum_{i=1}^{m} \sum_{j=1}^{m} c_{ij}^2 \|x - y\|^2, \end{aligned} \tag{30.16}$$

and so

$$q_T^2 \leqslant \bar{q}_T^2 = \sum_{i=1}^{m} \sum_{j=1}^{m} c_{ij}^2.$$

Similarly,

$$q_{PT}^2 \leqslant \bar{q}_{PT}^2 = \sum_{i=1}^{N} \sum_{j=1}^{m} c_{ij}^2, \qquad q_{QT}^2 \leqslant \bar{q}_{QT}^2 = \sum_{i=N+1}^{m} \sum_{j=1}^{m} c_{ij}^2,$$

$$q_{TP}^2 \leqslant \bar{q}_{TP}^2 = \sum_{i=1}^{m} \sum_{j=1}^{N} c_{ij}^2, \qquad q_{TQ}^2 \leqslant \bar{q}_{TQ}^2 = \sum_{i=1}^{m} \sum_{j=N+1}^{m} c_{ij}^2,$$

$$q_{PTP}^2 \leqslant \bar{q}_{PTP}^2 = \sum_{i=1}^{N} \sum_{j=1}^{N} c_{ij}^2, \qquad q_{QTP}^2 \leqslant \bar{q}_{QTP}^2 = \sum_{i=N+1}^{m} \sum_{j=1}^{N} c_{ij}^2,$$

$$q_{PTQ}^2 \leqslant \bar{q}_{PTQ}^2 = \sum_{i=1}^{N} \sum_{j=N+1}^{m} c_{ij}^2, \qquad q_{QTQ}^2 \leqslant \bar{q}_{QTQ}^2 = \sum_{i=N+1}^{m} \sum_{j=N+1}^{m} c_{ij}^2.$$

Since l_2 is a Hilbert space, we may use the results of §§16–18.

Finally, we consider the case

$$\varphi_i(x_1, x_2, \ldots, x_m) = \sum_{j=1}^{m} a_{ij} f_j(x_j) + b_i, \tag{30.17}$$

where the a_{ij} and b_i are numbers and $f_j(x_j)$ functions such that

$$m \leqslant m_j \leqslant \frac{f_j(x_j) - f_j(y_j)}{x_j - y_j} \leqslant M_j \leqslant M, \tag{30.18}$$

where m, m_j, M and M_j are certain numbers.

Set $\mu = (m + M)/2$, and suppose that the matrix $\{a_{ij}\}$ is symmetric. We may then use the results of §21; the operator A is represented by the matrix $A = \{a_{ij}\}$, and $f = \{f_i\}$ and $\varphi = \{b_i\}$. The underlying space may again be taken as l_2.

Proceeding as in §21, we can establish a sufficient condition for system (30.2) to have a unique solution.

THEOREM 30.1. *If*

$$\frac{M - m}{2} < \rho(S_A, \mu), \tag{30.19}$$

where $\rho(S_A, \mu)$ is the distance from the point μ to the spectrum of the matrix A, then system (30.2) *has a unique solution in l_2.*

No difficulties should be met in considering the more general case, in which the conditions are imposed not in the entire space but only in some subset.

REMARK. If system (30.2) is infinite, all the conditions involved are analogous to those in the finite-dimensional case, except that the finite sums Σ_1^m are replaced by infinite series Σ_1^∞, which are of course assumed to be convergent.

We now turn to a few numerical examples.

EXAMPLE 1. We consider a simple countable system of linear algebraic equations

$$x_i = b_i + \lambda \sum_{j=1}^{\infty} \frac{1}{2^{i+j}} x_j \qquad (i = 1, 2, \ldots), \tag{30.20}$$

with the readily determined exact solution

$$x_i = b_i + \frac{3\lambda}{3-\lambda} \cdot \frac{1}{2^i} \sum_{j=1}^{\infty} \frac{b_j}{2^j}.$$

In this case, E is the space of number sequences $x = \{x_i\}$ $(i = 1, 2, \ldots)$. If the projection P is defined to be P_N, determination of the vectors $x_n = \{x_{i,n}\}$ by the method (9.4) involves the system

$$x_{i,n} = b_i + \lambda \sum_{j=1}^{\infty} \frac{1}{2^{i+j}} x_{j,n} \qquad (i = 1, 2, \ldots, N),$$

$$x_{i,n} = b_i + \lambda \sum_{s=0}^{k-2} \left(\frac{\lambda}{3 \cdot 4^N}\right)^s \sum_{j=1}^{\infty} \frac{b_j}{2^{i+j}} + \frac{\lambda^2}{3}\left(1 - \frac{1}{4^N}\right) \sum_{s=0}^{k-2} \left(\frac{\lambda}{3 \cdot 4^N}\right)^s \sum_{j=1}^{\infty} \frac{x_{j,n}}{2^{i+j}}$$

$$+ \frac{\lambda^k}{(3 \cdot 4^N)^{k-1}} \sum_{j=1}^{\infty} \frac{x_{j,n-1}}{2^{i+j}} \qquad (i = N+1, \ldots). \tag{30.21}$$

If the algorithm employed is (9.7), we have a system for the vectors $x_n = \{x_{i,n}\}$:

$$x_{i,n} = b_i + \lambda \sum_{s=0}^{k-2} \left(\frac{\lambda}{3 \cdot 4^N}\right)^s \sum_{j=N+1}^{\infty} \frac{b_i}{2^{i+j}} + \lambda \sum_{s=0}^{k-1} \left(\frac{\lambda}{3 \cdot 4^N}\right)^s \sum_{j=1}^{N} \frac{x_{j,n}}{2^{i+j}}$$

$$+ \frac{\lambda^k}{(3 \cdot 4^N)^{k-1}} \sum_{i=N+1}^{\infty} \frac{x_{j,n-1}}{2^{i+j}} \qquad (i = 1, 2, \ldots). \tag{30.22}$$

The usual method of successive approximations, applied to system (30.20) for arbitrary b_i, is convergent only if $|\lambda| < 3$, whereas algorithms (30.21) and (30.22) are convergent for all λ such that

$$\left| \frac{\left(\dfrac{\lambda}{3 \cdot 4^N}\right)^k}{1 - \dfrac{\lambda}{3}\left(1 - \dfrac{1}{4^N}\right) \dfrac{1 - \left(\dfrac{\lambda}{3 \cdot 4^N}\right)^k}{1 - \dfrac{\lambda}{3 \cdot 4^N}}} \right| < 1.$$

This inequality is clearly valid for $\lambda \neq 3$ and sufficiently large N.

EXAMPLE 2. The system of nonlinear algebraic equations

$$x_i = \frac{1}{2} - \frac{1}{4^i} + \sum_{j=1}^{\infty} \frac{1}{2^{i+j}} x_j^i \quad (i = 1, 2, \ldots) \tag{30.23}$$

has the solution $x_i = ½$ $(i = 1, 2, \ldots)$. In this case,

$$Tx = \left\{ \frac{1}{2} - \frac{1}{4^i} + \sum_{j=1}^{\infty} \frac{1}{2^{\,i+j}} x_j^i \right\}.$$

For the underlying space we take the space of m-bounded number sequences. For $N = 2$, this gives

$$P_N Tx = \begin{cases} \frac{1}{2} - \frac{1}{4^i} + \sum_{j=1}^{\infty} \frac{1}{2^{i+j}} x_j^i & (i = 1, 2), \\ 0 \quad (i = 3, 4, \ldots), \end{cases}$$

$$Q_N Tx = \begin{cases} 0 \quad (i = 1, 2), \\ \frac{1}{2} - \frac{1}{4^i} + \sum_{j=1}^{\infty} \frac{1}{2^{i+j}} x_j^i & (i = 3, 4, \ldots). \end{cases}$$

Applying Sokolov's method $x_n = P_n Tx_n + Q_n Tx_{n-1}$, with initial approximation $x_{i,0} = \{0\}$, we obtain the following system for the first approximation:

$$x_{1,1} = \frac{1}{4} + \sum_{j=1}^{\infty} \frac{1}{2^{1+j}} x_{j,1}, \qquad x_{2,1} = \frac{7}{16} + \sum_{j=1}^{\infty} \frac{1}{2^{2+j}} x_{j,1}^2,$$

$$x_{i,1} = \frac{1}{2} - \frac{1}{4^i} \qquad (i = 3, 4, \ldots). \tag{30.24}$$

Solving, we find one of the solutions:

$$x_{1,1} = 0.499; \quad x_{2,1} = 0.500; \quad x_{i,1} = \frac{1}{2} - \frac{1}{4^i} \quad (i = 3, 4, \ldots).$$

The absolute error of the first approximation is $\|x - x_1\| = 0.016$, the relative error is $\|x - x_1\| / \|x\| = 3.2\%$.

Now, in the ball $R(\|x\| \leqslant M \leqslant ½)$ we have the estimate

$$\|Q_N Ty + P_N Tx\| \leqslant F(M) = \frac{1}{2},$$

and so $M = M^* = ½$. The bounds for Lipschitz constants in this ball are

$$\bar{q}_T = \frac{1}{2}; \; \bar{q}_{QT} = \frac{3}{8} M^2 = \frac{3}{32}; \; \bar{q}_{PTP} = \frac{3}{8}; \; \bar{q}_{PTQ} = \frac{1}{8};$$

$$\bar{q}_{QTP} = \frac{9}{32} M^2 = \frac{9}{128}; \; \bar{q}_{QTQ} = \frac{3}{32} M^2 = \frac{3}{128}$$

and

$$\varepsilon_1 = \bar{q}_{QTQ} + \bar{q}_{PTQ}\, \bar{q}_{QTP} (1 - \bar{q}_{PTP})^{-1} = \frac{3}{20} M^2 = \frac{3}{80}.$$

EXAMPLE 3. Consider a system of type (30.2) with the functions φ_i defined by formulas of type (30.17):

$$\begin{aligned} x_1 &= 2\left(4x_2 + \frac{1}{2}\sin x_2\right) + 1, \\ x_2 &= 2\left(4x_1 - \frac{1}{4}\sin x_1\right) - 3\left(4x_2 + \frac{1}{2}\sin x_2\right) - 1. \end{aligned} \tag{30.25}$$

In this case,

$$A = \{a_{ij}\} = \begin{pmatrix} 0 & 2 \\ 2 & -3 \end{pmatrix}, \quad b_1 = 1, \; b_2 = -1,$$

$$f_1(x_1) = 4x_1 - \frac{1}{4}\sin x_1, \qquad f_2(x_2) = 4x_2 + \frac{1}{2}\sin x_2.$$

Since $m_1 = 3.75, M_1 = 4.25$ and $m_2 = 3.5, M_2 = 4.5$, we have $m = 3.5, M = 4.5$ and $\mu = 4$.

Since the characteristic values of the matrix A are 1 and $-\frac{1}{4}$, it follows that $\rho(S_A, \mu) = 3$, and so the sufficient condition for the existence of a unique solution is satisfied ($(M - m)/2 = 0.5 < 3 = \rho(S_A, \mu)$).

However, the contraction mapping principle is not applicable to this system, since the operator in question is not contractive.

EXAMPLE 4. Consider the infinite system of linear algebraic equations

$$x_i = a_i + \frac{1}{2}\sum_{j=1}^{\infty} \frac{1}{(2i)^j} x_j \tag{30.26}$$

in the sequence space normed by setting $\|x\| = \sup_{i=1,2,\dots} |x_i|$ ($x = \{x_i\}$).

According to algorithm (17.5) for $k = 1$, the successive approximations $x_{i,n}$ to the solution of (30.26) are solutions of the systems

$$\begin{aligned} x_{i,n} &= a_i + \frac{1}{2}\sum_{j=1}^{\infty} \frac{1}{(2i)^j} x_{j,n} \quad (i = 1, 2, \dots, N), \\ x_{i,n} &= a_i + \frac{1}{2}\sum_{j=1}^{N} \frac{1}{(2i)^j} x_{j,n} + \sum_{j=N+1}^{\infty} \frac{1}{(2i)^j} x_{j,n-1} \quad (i = N+1, \dots), \end{aligned} \tag{30.27}$$

where $\{x_{i,0}\} \in m$. Here we have the following estimates:

$$q_{P_N T} \leqslant \frac{1}{2}; \; q_{QNT} \leqslant \frac{1}{2(2N+1)}; \; q_{P_N T P_N} \leqslant 1 - \frac{1}{2^N};$$

$$q_{P_N T Q_N} \leqslant \frac{1}{2^{N+1}}; \; q_{Q_N T P_N} \leqslant \frac{2^N (N+1)^N - 1}{2^{N+1}(N+1)^N (2N+1)};$$

$$q_{Q_N T Q_N} \leqslant \frac{1}{2^{N+1}(N+1)^N (2N+1)}.$$

§31. Integral equations

We shall now consider some applications of the general theory to approximate solution of integral equations of the types

$$x(t) = \varphi(t) + \int_0^1 K(t, s) f(t, s, x(s))\, ds, \tag{31.1}$$

and

$$x(t) = \varphi(t) + \int_0^t H(t, s) f(t, s, x(s))\, ds. \tag{31.2}$$

Here $\varphi(t)$, $K(t, s)$ and $f(t, s, x)$ are prescribed functions, and $x(t)$ is the unknown function.

If $f(t, s, x) = a(t, s)x + b(t, s)$, the equations (31.1) and (31.2) are linear.

It is clear that (31.2) is a special case of the general equation (31.1) obtained by setting

$$K(t, s) = \begin{cases} H(t, s) & \text{for } 0 \leqslant s \leqslant t, \\ 0 & \text{for } t \leqslant s \leqslant 1. \end{cases}$$

The derivation of convergence criteria and error estimates for the various approximate methods depends here to a considerable degree on the specific function space in which the equations are considered. We shall devote attention to only some of the function spaces, and only to a few of the solution methods studied in the previous chapters.

1. We first consider the general error estimate established in §2.

For equation (31.1), the operator T is defined by

$$Tx = \varphi(t) + \int_0^1 K(t, s) f(t, s, x(s))\, ds. \tag{31.3}$$

Let us assume that in a neighborhood D of a point y the function $f(t, s, x)$ may be expanded in powers of $x - y$:

$$f(t, s, x) = f(t, s, y) + \frac{1}{1!} f'_x(t, s, y)(x - y) + \frac{1}{2!} f''_x(t, s, y)(x - y)^2 + \dots$$
$$+ \frac{1}{n!} f_x^{(n)}(t, s, y)(x - y)^n + \frac{1}{(n+1)!} f_x^{(n+1)}(t, s, \theta)(x - y)^{n+1} \tag{31.4}$$

(θ is some point in $[y, x]$).

If the norm is defined by

$$\| x \| = | x(t) |, \tag{31.5}$$

we can define the operator $a = V$ (see §2) as

$$V\rho = \int_0^1 |K(t, s)| \Big[|f'_x(t, s, y)| \rho(s) + \frac{1}{2!} |f''_x(t, s, y)| \rho^2(s) + \dots \tag{31.6}$$
$$+ \frac{1}{n!} |f_x^{(n)}(t, s, y)| \rho^n(s) + \frac{1}{(n+1)!} \sup_{\theta \in D} |f_x^{(n+1)}(t, s, \theta)| \rho^{n+1}(s) \Big] ds.$$

If the underlying space is the space C of continuous functions, and we assume that all the functions figuring in (31.1) are such that the operator T is defined in this space (for example, the functions are continuous in their domains of definition), we obtain the following expression for V:

$$V\rho = \sum_{i=1}^{n} \frac{1}{i!} L_y^{(i)} \rho^i + \frac{1}{(n+1)!} L f^{(n+1)} \rho^{n+1}, \tag{31.7}$$

where

$$L_y^{(i)} = \sup_t \int_0^1 |K(t, s)| \, |f_x^{(i)}(t, s, y(s))| \, ds;$$

$$L = \sup_t \int_0^1 |K(t, s)| \, ds, \quad f^{(n+1)} = \sup_{\substack{t, s \in [0,1] \\ \theta \in D}} |f_x^{(n+1)}(t, s, \theta)|.$$

Treating (31.1) as a nonlinear operator equation in the space L_2 of square-integrable functions, we have the following expression for V:

$$V\rho = \sum_{i=1}^{n} \frac{1}{i!} \overline{L}^{(n+1)} \rho^i + \frac{1}{(n+1)!} \overline{L}^{(n+1)} \rho^{n+1}, \tag{31.8}$$

where

$$\overline{L}_y^{(i)} = \sqrt{\int_0^1 \int_0^1 |K(t, s) f_x^{(i)}(t, s, y(s))|^2 dt ds};$$

$$\overline{L}^{(n+1)} = \sqrt{\int_0^1 \int_0^1 |K(t, s) f^{(n+1)}(t, s)|^2 \, dt ds};$$

$$f^{(n+1)}(t, s) \geqslant |f_x^{(n+1)}(t, s, \theta)|.$$

We now turn to (31.2). In this case the operator T is

$$Tx = \varphi(t) + \int_0^t K(t, s) f(t, s, x(s)) \, ds. \tag{31.9}$$

Proceeding as we did for the operator (31.3), we can derive expressions for the operator V for equation (31.2) in various function spaces. We shall only consider the case that the function $f(t, s, x)$ satisfies

$$|f(t, s, x) - f(t, s, y)| \leqslant C(t, s) |x - y| \quad (-\infty < x, y < \infty). \tag{31.10}$$

We may then write V as

$$V\rho = \int_0 |H(t, s)| C(t, s) \rho(s) \, ds. \tag{31.11}$$

If $|H(t, s)|C(t, s) \leqslant K(t)L(s)$, we can estimate $|x^*(t) - v(t)|$, where $x^*(t)$ is a solution of (31.2) and $v(t)$ any function in the space:

$$|x^*(t) - v(t)| \leqslant |\varepsilon(t)| + K(t)\int_0^t L(s)|\varepsilon(s)| e^{\int_s^t K(\tau)L(\tau)d\tau} ds, \quad (31.12)$$

where

$$\varepsilon(t) = \varphi(t) + \int_0^t H(t, s) f(t, s, v(s))\, ds - v(t).$$

To prove (31.12), it is sufficient to note that $x^*(t) - v(t)$ satisfies the inequality

$$|x^*(t) - v(t)| \leqslant |\varepsilon(t)| + \int_0^t K(t) L(s) |x^*(s) - v(s)|\, ds,$$

and the expression on the right of (31.12) is a solution of the equation

$$x(t) = |\varepsilon(t)| + \int_0^t K(t) L(s) x(s)\, ds \quad (31.13)$$

(see §2).

EXAMPLE 5. The solution of the equation

$$x(t) = \frac{1}{6}\pi - 2\left(\frac{1}{3}\pi - 1\right)(t - t^2) + 8\int_0^1 G(t, s)[x(s) - \sin x(s)]\, ds,$$

$$G(t, s) = \begin{cases} s(1-t) & (0 \leqslant s \leqslant t), \\ t(1-s) & (t \leqslant s \leqslant 1) \end{cases} \quad (31.14)$$

is $x(t) = \pi/6$.

Treating (31.14) as a nonlinear operator equation in the space C of continuous functions on $[0, 1]$, we shall find an estimate for the norm of the solution. Here

$$\varphi(t) = \frac{1}{6}\pi - 2\left(\frac{1}{3}\pi - 1\right)(t - t^2); \; K(t, s) = 8G(t, s);$$

$$f(t, s, x) = x - \sin x.$$

Since

$$f'_x(t, s, 0) = 1 - \cos 0 = 0; \; f''_x(t, s, 0) = \sin 0 = 0; \; f'''_x(t, s, x) = \cos x,$$

and

$$L = \sup_t 8\int_0^1 G(t, s)\, ds = \sup_t 8 \cdot \frac{1}{2}(t - t^2) = 1,$$

$$f^{(3)} = \sup_{\theta \in (-\infty, \infty)} |\cos\theta| = 1, \; L_y^{(1)} = L_y^{(2)} = 0,$$

it follows that $V\rho = \rho^3/6$. Since

$$\|\varepsilon(0)\| = \sup_t \left| \frac{1}{6}\pi - 2\left(\frac{1}{3}\pi - 1\right)(t - t^2) \right| = \frac{1}{6}\pi,$$

we can set up an equation from which the norm of the solution may be estimated:

$$\frac{1}{6}\pi + \frac{1}{6} M^3 = M, \tag{31.15}$$

and this equation has a unique solution on the positive semiaxis: $M \approx 0.55$.

Thus $\|x(t)\| \leqslant 0.55$; the exact solution is $\|x(t)\| = \pi/6$.

2. Suppose that the function $f(t, s, x)$ in (31.1) is independent of t; that is, $f(t, s, x) = f(s, x)$. Then (31.1) is an equation of Hammerstein type:

$$x(t) = \varphi(t) + \int_0^1 K(t, s) f(s, x(s))\, ds. \tag{31.16}$$

We apply the algorithm of §7 to solve this equation. The operator T is defined by

$$Tx = \varphi(t) + \int_0^1 K(t, s) f(s, x(s))\, ds. \tag{31.17}$$

As operators S_n "approximating" T we take the operators defined by

$$S_n x = \varphi(t) + \int_0^1 K_n(t, s) f(s, x(s))\, ds, \tag{31.18}$$

where $K_n(t, s)$ are singular kernels approximating $K(t, s)$, e.g. partial sums of the Fourier series of $K(t, s)$ or of its Taylor series.

Assuming that $K(t, s)$ and $K_n(t, s)$ are continuous on $[0, 1]$ in both variables, and the function $f(s, x)$ satisfies the condition

$$|f(s, x) - f(s, y)| \leqslant C(s)|x - y| \quad (-\infty < x, y < \infty), \tag{31.19}$$

we shall consider equation (31.16) in the function space normed by (31.5). As a Lipschitz operator for T, we can take

$$L_T\rho = \int_0^1 |K(t, s)| C(s) \rho(s)\, ds, \tag{31.20}$$

and the Lipschitz operators for the operators S_n and $T - S_n$ will be

$$L_{S_n}\rho = \int_0^1 |K_n(t, s)| C(s) \rho(s)\, ds, \tag{31.21}$$

$$L_{T-S_n}\rho = \int_0^1 |K(t, s) - K_n(t, s)| C(s) \rho(s)\, ds. \tag{31.22}$$

Lipschitz operators for the other operators involved are also readily found.

We now assume that (31.19) holds not in the entire space but only in some ball $\|x(t) - \varphi(t)\| \leqslant M(t)$, in which

$$|f(s, x(s))| \leqslant F(M(s)) \tag{31.23}$$

and

$$|K(t, s) - K_n(t, s)| \leqslant H_{T-S}(t, s), \quad |K_n(t, s)| \leqslant H_S(t, s). \tag{31.24}$$

Then

$$\|S_n x + (T - S_n) y - \varphi\| \leqslant F_S(M) + F_{T-S}(M), \tag{31.25}$$

where

$$F_{T-S}(M) = \int_0^1 H_{T-S}(t, s) F(M(s))\, ds \tag{31.26}$$

and

$$F_S(M) = \int_0^1 H_S(t, s) F(M(s))\, ds. \tag{31.27}$$

On the other hand, if we deal with equation (31.16) in some Banach space, the Lipschitz operators become Lipschitz constants, and the radius of the ball in which convergence is guaranteed will be determined by a scalar inequality.

For example, consider the Banach space with norm

$$\|x\| = \sup_{t \in [0,1]} |w(t) x(t)|, \tag{31.28}$$

where $w(t)$ is some function positive on $[0, 1]$.

For simplicity's sake, we consider algorithm (8.2) with $k = 1$. The successive approximations $x_n(t)$ are defined by

$$x_n(t) = \varphi(t) + \int_0^1 [K(t, s) - K_S(t, s)] f(s, x_{n-1}(s))\, ds + \int_0^1 K_S(t, s) f(s, x_n(s))\, ds, \tag{31.29}$$

where $x_0(t)$ is an arbitrary function in the space and $K_S(t, s)$ a singular kernel approximating $K(t, s)$ (for example, a partial sum of the Fourier or Taylor series of $K(t, s)$).

Then $F(x, y) = Rx + Sy$, where

$$Rx = \varphi(t) + \int_0^1 K_S(t, s) f(s, x(s))\, ds,$$

$$Sx = \int_0^1 [K(t, s) - K_S(t, s)] f(s, x(s))\, ds.$$

Algorithm (11.3) may be written as a system of equations

$$u_n(t) = \varphi(t) + \int_0^1 K_S(t, s) f[s, u_n(s) + v_n(s)]\, ds,$$
$$v_n(t) = \int_0^1 [K(t, s) - K_S(t, s)] f[s, u_n(s) + v_{n-1}(s)]\, ds, \tag{31.30}$$

setting $k = 1$,

$$Ux = \varphi(t) + \int_0^1 K_S(t, s) f(s, x(s))\, ds,$$

$$Sx = \int_0^1 [K(t, s) - K_S(t, s)] f(s, x(s))\, ds.$$

Let us estimate the Lipschitz constant for T. We have

$$Tx - Ty = \int_0^1 K(t, s) [f(s, x) - f(s, y)]\, ds,$$

and by (2.28)

$$\begin{aligned} \|Tx - Ty\| &= \sup_t \left| \int_0^1 w(t) K(t, s) [f(s, x) - f(s, y)]\, ds \right| \\ &\leqslant \sup_t \int_0^1 w(t) \,|K(t, s)|\, C(s) \frac{1}{w(s)} w(s) \,|x(s) - y(s)|\, ds| \\ &\leqslant \sup_t \int_0^1 w(t)\, |K(t, s)|\, C(s) \frac{1}{w(s)}\, ds \cdot \sup_t w(t)\, |x(t) - y(t)| \\ &= \sup_t \int_0^1 w(t)\, |K(t, s)|\, C(s) \frac{1}{w(s)}\, ds \cdot \|x - y\|. \end{aligned} \tag{31.31}$$

Consequently, we have the following inequality for the Lipschitz constant q_T:

$$q_T \leqslant \sup_t \int_0^1 w(t)\, |K(t, s)|\, C(s) \frac{1}{w(s)}\, ds.$$

The Lipschitz constants for S and T–S are found similarly:

$$q_S \leqslant \sup_t \int_0^1 w(t)\, |K_S(t, s)|\, C(s) \frac{1}{w(s)}\, ds,$$

$$q_{T-S} \leqslant \sup_t \int_0^1 w(t)\, |K(t, s) - K_S(t, s)|\, C(s) \frac{1}{w(s)}\, ds.$$

The convergence criterion for the process (31.29) may be written

$$q_{T-S} + q_S < 1. \tag{31.32}$$

Using the Cauchy-Bunjakovskiĭ-Schwarz inequality for integrals, we can slightly weaken this criterion. Let equation (31.16) have a solution $x^*(t)$. Setting

$\Delta_n(t) = x^*(t) - x_n(t)$, we ueduce from (31.29) that

$$\|\Delta_n(t)\| \leqslant \sup_t \int_0^1 w(t)\,|K(t, s) - K_S(t, s)|\,C(s)\frac{1}{w(s)}\,w(s)\,|\Delta_{n-1}(s)|\,ds$$
$$+ \sup_t \int_0^1 w(t)\,|K_S(t, s)|\,C(s)\frac{1}{w(s)}\,w(s)\,|\Delta_n(s)|\,ds. \tag{31.33}$$

Hence, applying the Cauchy-Bunjakovskiĭ-Schwarz inequality to the first term on the right of (31.33) and estimating the second term, we obtain

$$\|\Delta_n\| \leqslant \bar{q}_{T-S}\sqrt{|w(s)\,\Delta_{n-1}(s)|^2\,ds} + q_S\|\Delta_n\|,$$
$$\|\Delta_n\| \leqslant \bar{q}_{T-S}(1-q_S)^{-1}\sqrt{\int_0^1 |w(s)\,\Delta_{n-1}(s)|^2 ds}, \tag{31.34}$$

where

$$\bar{q}_{T-S} = \sup_t \sqrt{\int_0^1 \left[w(t)\,|K(t, s)|\,C(s)\frac{1}{w(s)}\right]^2 ds}\,.$$

Next, starting from the same equality (31.29) and applying the Cauchy-Bunjakovskiĭ-Schwarz inequality, we get

$$\sqrt{\int_0^1 |w(t)\,\Delta_n(t)|^2 dt} \leqslant \tilde{q}_{T-S}\sqrt{\int_0^1 |w(t)\,\Delta_{n-1}(t)|^2 dt}$$
$$+ \tilde{q}_S\sqrt{\int_0^1 |w(t)\,\Delta_n(t)|^2 dt,}$$

whence

$$\sqrt{\int_0^1 |w(t)\,\Delta_n(t)|^2 dt} \leqslant \frac{\tilde{q}_{T-S}}{1-\tilde{q}_S}\sqrt{\int_0^1 |w(t)\,\Delta_{n-1}(t)|^2 dt}\,, \tag{31.35}$$

where

$$\tilde{q}_s = \sqrt{\int_0^1\int_0^1 \left[w(t)\,|K_S(t, s)|\,C(s)\frac{1}{w(s)}\right]^2 dtds,}$$

$$\tilde{q}_{T-S} = \sqrt{\int_0^1\int_0^1 \left[w(t)\,|K(t, s) - K_S(t, s)|\,C(s)\frac{1}{w(s)}\right]^2 dtds}\,.$$

Repeating the procedure, we find that

$$\sqrt{\int_0^1 |w(t)\,\Delta_n(t)|^2 dt} \leqslant \tilde{\varepsilon}^{n-p}\sqrt{\int_0^1 |w(t)\,\Delta_{p-1}(t)|^2 dt} \tag{31.36}$$
$$(1 \leqslant p \leqslant n).$$

Consequently the convergence criterion for algorithm (31.29) becomes $\epsilon < 1$ ($\widetilde{q}_S < 1$), where $\epsilon = \widetilde{q}_{T-S}/(1-\widetilde{q}_S)$.

We now treat (31.16) as a nonlinear operator equation in a Hilbert space H with norm

$$\|x\| = \sqrt{\int_0^1 |w(t)\,x(t)|^2 dt}\,. \tag{31.37}$$

Here we must assume that the functions $\varphi(t)$ and $K(t, s)$ are square-integrable in the respective domains $t \in [0, 1]$ and $t, s \in [0, 1]$, and the function $f(t, x)$ satisfies Carathéodory's condition, i.e., is measurable in t for each fixed x and continuous in x for almost all $t \in [0, 1]$.

Let $\{\varphi_i(t)\}$ be some complete orthonormal system of linearly independent functions on $[0, 1]$:

$$\int_0^1 w^2(t)\,\varphi_i(t)\,\varphi_j(t)\,dt = \delta_{ij},$$

where $\delta_{i,j}$ is the Kronecker delta.

Consider the projection P' of H onto the subspace $H_{P'}$ spanned by the first k elements of the basis $\{\varphi_i(t)\}$:

$$P'x(t) = \sum_{i=1}^{k} (x, \varphi_i)\,\varphi_i, \tag{31.38}$$

where (x, y) denotes the scalar product of elements $x, y \in H$:

$$(x, y) = \int_0^1 w^2(t)\,x(t)\,y(t)\,dt.$$

Since P' is an orthogonal projection, it satisfies (16.1).

Applied to equation (31.16), algorithm (7.13) may be written

$$x_n(t) = \varphi(t) + \int_0^t [K(t, s) - K_S(t, s)]\,f[s, Q'x_{n-1}(s) + P'x_n(s)]\,ds$$
$$+ \int_0^1 K_S(t, s)\,f[s, x_n(s)]\,ds, \tag{31.39}$$

where P' is the orthogonal projection just defined, $Q' = I - P'$ (where I is the identity operator), and $K_S(t, s)$ is defined by

$$K_S(t, s) = \sum_{i=1}^{l} \int_0^1 K(t, s)\,\psi_i(t)\,dt \cdot \psi_i(t)$$

($\{\psi_i(t)\}$, like $\{\varphi_i(t)\}$, is a complete orthonormal system of linearly independent functions on $[0, 1]$).

Using the Cauchy-Bunjakovskiĭ-Schwarz inequality, we can estimate the Lipschitz constants for the various operators involved. For example, for the Lipschitz operator $q_{QQ'T}$ we have

$$q_{QQ'T} \leqslant \sqrt{\int_0^1\int_0^1 \left[\sum_{i=1}^{k} \int_0^1 [K(\tau, s) - K_S(\tau, s)]\, \varphi_i(\tau)\, d\tau \varphi_i(t)\right]^2 C^2(s)\, dt ds}\,.$$

The estimates for the other operators are similar.

We now consider some numerical examples.

EXAMPLE 6. One of the solutions of the integral equation

$$x(t) = \frac{1}{5} t + \frac{5}{6} t^2 + \int_0^1 t(ts - 1)\, x^2(s)\, ds \qquad (31.40)$$

is the function $x(t) = t^2$.

This equation may be regarded as a special case of the operator equation $x = Tx$ in the space L_2 of square-integrable functions on $[0, 1]$:

$$Tx = \frac{1}{5} t + \frac{5}{6} t^2 + \int_0^1 t(ts - 1)\, x^2(s)\, ds.$$

We let P be the orthogonal projection defined by

$$Px = 3t \int_0^1 sx(s)\, ds \quad (\varphi_1 = \sqrt{3}t).$$

We shall use algorithm (31.39) to solve equation (31.40) in the following two special cases:

a) $P' = 0,\ \psi_1 = \sqrt{3}t$;

b) $P' = P,\ K_S(t, s) = 0\ (\psi_1 = 0)$.

In case (a), the successive approximations $u_n(t)$ to the solution of equation (31.40) are defined by

$$u_n(t) = \frac{1}{5} t + \frac{5}{6} t^2 + \int_0^1 t\left(\frac{3}{4} s - 1\right) u_n^2(s)\, ds$$
$$+ \int_0^1 t\left(t - \frac{3}{4}\right) s u_{n-1}^2(s)\, ds \quad (n = 1, 2, \ldots). \qquad (31.41)$$

In case (b), we have the following equations for $v_n(t)$:

$$v_n(t) = \frac{1}{5} t + \frac{5}{6} t^2 + \int_0^1 t(ts - 1)\, [v_{n-1}(s) + \alpha_n(s)]^2\, ds, \qquad (31.42)$$

where $\alpha_n(t) = 3t \int_0^1 s[v_n(s) - v_{n-1}(s)]\, ds$.

Setting $u_0 = v_0 = t/5 + 5t^2/6$ and solving equations (31.41) and (31.42),

t	x	$\eta_0 = u_0 = v_0$	$x - \eta_0$	η_1	$x - \eta_1$	u_1	$x - u_1$	v_1	$x - v_1$	y	$x - y$
0,0	0,00	0,0000	0,0000	0,0000	0,0000	0,0000	0,0000	0,0000	0,0000	0,0000	0,0000
0,1	0,01	0,0283	−0,0183	0,0067	0,0033	0,0083	0,0017	0,0103	−0,0003	0,0744	−0,0644
0,2	0,04	0,0733	−0,0333	0,0339	0,0061	0,0372	0,0028	0,0406	−0,0006	0,1488	−0,1088
0,3	0,09	0,1350	−0,0450	0,0816	0,0084	0,0865	0,0035	0,0907	−0,0007	0,2233	−0,1333
0,4	0,16	0,2133	−0,0533	0,1499	0,0101	0,1564	0,0036	0,1607	−0,0007	0,2977	−0,1377
0,5	0,25	0,3083	−0,0583	0,2387	0,0113	0,2468	0,0032	0,2506	−0,0006	0,3721	−0,1221
0,6	0,36	0,4200	−0,0600	0,3479	0,0121	0,3577	0,0023	0,3604	−0,0004	0,4465	−0,0865
0,7	0,49	0,5483	−0,0583	0,4777	0,0123	0,4891	0,0009	0,4901	−0,0001	0,5110	−0,0210
0,8	0,64	0,6933	−0,0533	0,6280	0,0120	0,6410	−0,0010	0,6397	0,0003	0,5954	0,0446
0,9	0,81	0,8550	−0,0450	0,7989	0,0111	0,8135	−0,0035	0,8091	0,0009	0,6698	0,1402
1,0	1,00	1,0333	−0,0333	0,9902	0,0098	1,0064	−0,0064	0,9985	0,0015	0,7442	0,2558

we find the approximatons:

$$u_1(t) = 1.02574t^2 - 0.01931t;$$
$$v_1(t) = 0.99458t^2 + 0.00390t.$$

Solving equation (31.40) by the usual method of successive approximations, with the same initial approximation $\eta_0 = t/5 + 5t^2/6$, we obtain the first approximation:

$$\eta_1(t) = 1.02574t^2 - 0.03556t.$$

If (31.40) is solved by the projective method with the same projection operator P, i.e., the Galerkin method with coordinate function $\varphi_1 = \sqrt{3}t$, the equation for the approximate solution $y(t)$ is

$$y(t) = \frac{33}{40}t + \int_0^1 t\left(\frac{3}{4}s - 1\right) y^2(s)\,ds, \tag{31.43}$$

whence $y(t) = 0.74423t$.

For the errors corresponding to these approximations, we have the following expressions:

$$x(t) - \eta_1(t) = 0.02574t^2 + 0.03556t;$$
$$x(t) - u_1(t) = -0.02574t^2 + 0.01931t;$$
$$x(t) - v_1(t) = 0.00542t^2 - 0.00390t;$$
$$x(t) - y(t) = t^2 - 0.74423t.$$

The absolute errors (in the L_2-norm) are

$$\|x(t) - \eta_1(t)\| = 0.0095;$$
$$\|x(t) - u_1(t)\| = 0.003;$$
$$\|x(t) - v_1(t)\| = 0.0006;$$
$$\|x(t) - y(t)\| = 0.11,$$

and the relative errors

$$\frac{\|x(t) - \eta_1(t)\|}{\|x(t)\|} = 2.1\,\%; \qquad \frac{\|x(t) - v_1(t)\|}{\|x(t)\|} = 0.14\,\%;$$

$$\frac{\|x(t) - u_1(t)\|}{\|x(t)\|} = 0.7\,\%; \qquad \frac{\|x(t) - y(t)\|}{\|x(t)\|} = 24.6\,\%.$$

The numerical values at a few points of $[0, 1]$, both of the exact solution to equation (31.40) and of the above approximate solutions, together with the error estimates, are given in the table,

EXAMPLE 7. Consider the simple linear integral equation

$$x(t) = t^2 + \int_0^1 (6 - ts)\, x(s)\, ds, \tag{31.44}$$

whose solution is

$$x(t) = t^2 - \frac{3}{62} t - \frac{23}{62}.$$

Let us solve this equation using algorithm (19.2) with $\varphi_{i,n} = 0$, $\psi_{1,n} = 1$ and $\psi_{i,n} = 0$ for $i > 1$; in other words, the successive approximations $x_n(t)$ are defined by

$$x_n(t) = t^2 + \int_0^1 (6 - ts)\, x_{n-1}(s)\, ds + D_n,$$

where the constants D_n are chosen so as to minimize the L_2-norm of the residual $x_n(t) - t^2 - \int_0^1 (6 - ts)x_n(s)\,ds$. Substituting the appropriate values of D_n in the equation for $x_n(t)$, we obtain a recurrence relation

$$x_n(t) = t^2 - \frac{107}{271} + \int_0^1 \left(\frac{401}{813} - t\right) s x_{n-1}(s)\, ds. \tag{31.45}$$

Thus this algorithm is equivalent to a classical successive approximation procedure applied to the new equation

$$x(t) = t^2 - \frac{107}{271} + \int_0^1 \left(\frac{401}{813} - t\right) s x(s)\, ds,$$

which is equivalent to the original equation (31.44). The advantage is that the process (31.45) is convergent for any initial approximation $x_0(t) \in L_2$, whereas the direct method of successive approximations is not applicable to the original equation.

Setting $x_0(t) = 0$, we obtain as a first approximation $x_1(t) = t^2 - 107/271$. Solving (31.44) by least squares (setting $y(t) = C$), we get $y(t) = -37/542$.

The maximum deviations of the errors $x(t) - x_1(t)$ and $x(t) - y(t)$ are 0.06 and 0.65, respectively.

EXAMPLE 8. In the space L_2, consider the linear integral equation

$$x(t) = 21 - 65t^2 + \int_0^1 \left[1 + \frac{1}{2}(t - s)^2\right] x(s)\, ds, \tag{31.46}$$

which is solved by the function $x(t) = 30 - 60t^2$.

Setting $f = 21 - 65t^2$ and $Ax = \int_0^1 [1 + (t - s)^2/2]\, ds$, we can write (31.46) as an operator equation $x = f + Ax$.

Let us apply algorithm (20.8) to equation (31.46), with $\varphi_1 = \varphi_1(t) = \psi_1 = \psi_1(t) = 1$. Then

$$PA(I-A)^{-1}f = \frac{65}{6},$$

$$PAx = \int_0^1 \left(1 + \frac{11}{60} - \frac{1}{2}s + \frac{1}{2}s^2\right) x(s)\, ds,$$

$$QAx = \int_0^1 \left[-\frac{11}{60} + \frac{1}{2}t^2 + \left(\frac{1}{2} - t\right)s\right] x(s)\, ds.$$

In this case the successive approximations (20.8) converge to a solution of equation (31.46), for the eigenvalue of QA of maximum absolute value is $-1/12$.

Equation (20.3) is now

$$x(t) = \frac{191}{6} - 65t^2 + \int_0^1 \left[-\frac{11}{60} + \frac{1}{2}t^2 + \left(\frac{1}{2} - t\right)s\right] x_0(s)\, ds. \qquad (31.47)$$

Taking $x_0 = 0$ as initial approximation, we find the first approximation

$$x_1(t) = \frac{191}{6} - 65t^2,$$

and the second approximation

$$x_2(t) = \frac{10\,729}{360} + \frac{1}{3}t - \frac{719}{12}t^2.$$

The errors of the first and second approximations are

$$\|x - x_1\| = \frac{3}{2}, \quad \|x - x_2\| = \frac{\sqrt{209}}{120}.$$

EXAMPLE 9. The nonlinear integral equation

$$x(t) = -\frac{1}{20}t + \int_0^1 \left[x^2(s) + \frac{1}{10}tsx(s)\right] ds \qquad (31.48)$$

has two easily found solutions:

$$x^*(t) = 1, \quad x^{**}(t) = \frac{3}{3541} - \frac{183}{3541}t.$$

We apply algorithm (8.2), setting

$$F(x, y) = -\frac{1}{20}t + \int_0^1 \left[x^2(s) + \frac{1}{10}tsy(s)\right] ds.$$

The operators $F_k(x, y)$ are then given by

$$F_k(x, y) = -\frac{3}{58}\left(1 - \frac{1}{30^k}\right)t + \int_0^1 \left\{\left[1 + \frac{3}{58}\left(1 - \frac{1}{30^{k-1}}\right)t\right] x^2(s) + \frac{1}{10\cdot 30^{k-1}} tsy(s)\right\} ds$$

and the successive approximations $x_n(t)$ are solutions of the integral equations

$$x_n(t) = -\frac{3}{58}\left(1-\frac{1}{30^k}\right)t + \int_0^1 \left\{\left[1+\frac{3}{58}\left(1-\frac{1}{30^{k-1}}\right)t\right]x_n^2(s) + \frac{1}{10\cdot 30^{k-1}}\, tsx_{n-1}(s)\right\} ds. \tag{31.49}$$

Taking $x_0(t) = 0$ as an initial approximation, we find two solutions of (31.49) in the first approximation; one of them tends to $x^*(t)$ as $k \to \infty$, the other to $x^{**}(t)$. The classical method of successive approximations is also applicable to equation (31.48), with initial approximation $x_0(t) = 0$, but it produces a sequence converging to the solution $x^{**}(t)$.

EXAMPLE 10. The linear integral equation

$$x(t) = 22 - 8t + \int_0^1 [36(t^2+s^2) - (1-2t)(1-2s)]\, x(s)\, ds \tag{31.50}$$

has a solution $x(t) = 3 - 6t$.

We define a projection P by $Px = \int_0^1 x(t)\,dt$. Since now $Tx = f + Ax$, where

$$f = 22 - 8t;\quad Ax = \int_0^1 36(t^2+s^2) - (1-2t)(1-2s)\, x(s)\, ds,$$

and the eigenvalues of the operator $(I - PA)^{-1}QA$ are greater than 1 in absolute value, it follows that algorithms (9.4) and (9.7) with $k = 1$ are divergent. Algorithm (9.10) with $k = 1$ is convergent for any initial approximation $x_0(t)$ in the space. According to this algorithm, the successive approximations $w_n(t)$ are determined by the equations

$$w_n(t) = 22 - 8t + 36\int_0^1 (t^2+s^2)\, w_n(s)\, ds - \int_0^1 (1-2t)(1-2s)\, w_{n-1}(s)\, ds. \tag{31.51}$$

Solving for $w_n(t)$, we get

$$w_n(t) = \frac{2434}{691} - 8t + \frac{1080}{691}t^2 + \int_0^1 \left[-\frac{1080}{691}t^2 + 2t - \frac{361}{691}\right](1-2s)\, w_{n-1}(s)\, ds,$$

whence we readily deduce that the algorithm is convergent (eigenvalue $=$ $-151/2073$).

EXAMPLE 11. In some function space E, consider the linear integral equation

$$x(t) = f(t) + \lambda\int_0^1 e^{ts}x(s)\, ds \tag{31.52}$$

($f(t) \in E$, and λ is a scalar parameter).

Denoting $f = f(t)$ and $Ax = \lambda \int_0^1 e^{ts}x(s)\,ds$, we may write equation (31.52) as $x = f + Ax$.

We shall solve (31.52) using algorithm (11.3) with $k = 1$, setting

$$Ux = f + \bar{A}x, \quad Sx = (A - \bar{A})\,x,$$

where $\bar{A}x = \lambda \int_0^1 x(s)\,ds$. For u_n and v_n we have systems of integral equations

$$\begin{aligned} u_n(t) &= f(t) + \lambda \int_0^1 u_n(s)\,ds + \lambda \int_0^1 v_n(s)\,ds, \\ v_n(t) &= \lambda \int_0^1 (e^{ts} - 1)\,u_n(s) + \lambda \int_0^1 (e^{ts} - 1)\,v_{n-1}(s)\,ds. \end{aligned} \tag{31.53}$$

When $\lambda = 1$ the algorithm is convergent and $x_n(t) = u_n(t) + v_n(t)$ converges to a solution of (31.52). In this case the algorithm $x_n = f + \bar{A}x_n + (A - \bar{A})x_{n-1}$ is useless when $\lambda = 1$.

When $\lambda = 1$ and $f(t) = e^t - (e^{t+1} - 1)/(t + 1)$, equation (31.52) has a solution $x(t) = e^t$. Setting $v_0(t) = 0$, we see from (31.53) that

$$x_1(t) = e^t - \frac{e^{t+1} - 1}{t+1} + \int_0^1 (e^{ts} - 1)\left(e^s - \frac{e^{s+1} - 1}{s+1}\right) ds + \frac{f_1}{1-a} \cdot \frac{e^t - 1}{t},$$

where

$$a = \int_0^1 \frac{e^t - 1}{t}\,dt, \quad f_1 = \int_0^1 \frac{e^t - 1}{t}\left(e^t - \frac{e^{t+1} - 1}{t+1}\right) dt.$$

§32. Finite and infinite systems of integral equations

In this section we apply some results from Chapters I and II to the solution of finite and infinite systems of integral equations

$$x_i(t) = \varphi_i(t) + \int_0^1 K_i(t, s)\, f_i\,[s, x_1(s), \ldots, x_m(s)]\,ds \qquad (i = 1, \ldots, m). \tag{32.1}$$

Here m is finite if system (32.1) is finite and infinite otherwise.

It is obvious that the system

$$x_i(t) = \varphi_i(t) + \int_0^t H_i(t, s)\, f_i\,[s, x_1(s), \ldots, x_m(s)]\,ds \qquad (i = 1, \ldots, m) \tag{32.2}$$

is a particular case of (32.1), obtained by putting

$$K_i(t, s) = \begin{cases} H_i(t, s) & \text{if } s \leqslant t, \\ 0 & \text{if } s \geqslant t. \end{cases}$$

We shall first assume that the functions $\varphi_i(t)$ and $K_i(t, s)$ are continuous for $t \in [0, 1]$ and $s \in [0, 1]$, and satisfy

$$|\varphi_i(t)| \leqslant \varphi_i < \varphi, \quad |K_i(t, s)| \leqslant K_i < K, \tag{32.3}$$

while the functions $f_i[s, x_1, \ldots, x_m]$ satisfy

$$|f_i(s, x_1, \ldots, x_m) - f_i(s, y_1, \ldots, y_m)| \leqslant \sum_{j=1}^{m} C_{ij}(s)\,|x_j - y_j| \quad (-\infty < x_j, y_j < \infty). \tag{32.4}$$

System (32.1) will be viewed as a nonlinear operator equation in the space E of m-normed sequences of functions $x(t) = \{x_i(t)\}$. The operator T is defined by $Tx = \{T_i x\}$, where

$$T_i x = \varphi_i(t) + \int_0^1 K_i(t, s) f_i[s, x_1, \ldots, x_m]\,ds. \tag{32.5}$$

As operators S_n approximating T we can take the operators $S_n = \{S_{ni}\}$ defined by

$$S_{ni} x = \varphi_i(t) + \int_0^1 K_{in}(t, s) f_i[s, x_1, \ldots, x_m]\,ds, \tag{32.6}$$

where $K_{in}(t, s)$ are kernels approximating $K_i(t, s)$, say partial sums of the Fourier or Taylor series of $K_i(t, s)$.

If the norm $\|x\|$ of an element $x = \{x_i(t)\}$ in E is defined by

$$\|x\| = \{|x_i(t)|\}, \tag{32.7}$$

in other words, the norm is the vector with components $|x_i(t)|$, then the Lipschitz operator L_T for T may be determined as follows:

$$\begin{aligned} \|Tx - Ty\| &= \left\{\left|\int_0^1 K_i(t, s)\,[f_i(s, x_1, \ldots, x_m) - f_i(s, y_1, \ldots, y_m)]ds\right|\right\} \\ &\leqslant \left\{\int_0^1 |K_i(t, s)| \sum_{j=1}^{m} C_{ij}(s)\,|x_j(s) - y_j(s)|\,ds\right\} \leqslant L_T\,\|x - y\|, \end{aligned} \tag{32.8}$$

so that

$$L_T = \{L_{T_i}\}, \qquad L_{T_i} x = \int_0^1 \sum_{j=1}^{m} |K_i(t, s)|\,C_{ij}(s)\,x_j(s)\,ds.$$

Lipschitz operators for the other operators are estimated in similar fashion. In particular, for the operators S_n and $T - S_n$ we have

$$\begin{aligned} L_{S_n} &= \{L_{S_{ni}}\}, \\ L_{S_{ni}} x &= \int_0^1 \sum_{j=1}^{m} |K_{in}(t, s)|\,C_{ij}(s)\,x_j(s)\,ds; \\ L_{T-S_n} &= \{L_{(T-S_n)i}\}, \\ L_{(T-S_n)i} x &= \int_0^1 \sum_{j=1}^{m} |K_i(t, s) - K_{in}(t, s)|\,C_{ij}(s)\,x_j(s)\,ds. \end{aligned}$$

The algorithm $x_n = S_n x_n + (T - S_n) x_{n-1}$ may be written

$$x_{i,n}(t) = \varphi_i(t) + \int_0^1 [K_i(t, s) - K_{in}(t, s)] f_i(s, x_{1,n-1}, \ldots, x_{m,n-1})\, ds$$
$$+ \int_0^1 K_{in}(t, s) f_i(s, x_{1,n}, \ldots, x_{m,n})\, ds. \tag{32.9}$$

If system (32.1) is finite, or if it is infinite and $\Sigma_{j=1}^{\infty} \int_0^1 C_{ij}(s)\, ds$ converges, we may approximate T by the operators

$$S_n = \begin{cases} S_{ni} & (i = 1, \ldots, N;\ N \leqslant m), \\ 0 & (i = N+1, \ldots, m), \end{cases}$$

where S_{ni}, $i = 1, \ldots, N$, is defined by (32.6). We then have

$$T - S_n = \{(T_i - S_{ni})\} \quad (i = 1, 2, \ldots, m),$$

$$(T_i - S_{ni})x = \begin{cases} \varphi_i(t) + \int_0^1 [K_i(t, s) - K_{in}(t, s)] f_i(s, x_1, \ldots, x_m)\, ds \\ \qquad (i = 1, \ldots, N), \\ \varphi_i(t) + \int_0^1 K_i(t, s) f_i(s, x_1, \ldots, x_m)\, ds \ (i = N+1, \ldots, m). \end{cases} \tag{32.10}$$

The algorithm $x_n = S_n x_n + (T - S_n)x_{n-1}$ is now written thus:

$$x_{i,n}(t) = \varphi_i(t) + \int_0^1 [K_i(t, s) - K_{in}(t, s)] f_i(s, x_{1,n-1}, \ldots, x_{m,n-1})\, ds$$
$$+ \int_0^1 K_{in}(t, s) f_i(s, x_{1,n}, \ldots, x_{m,n})\, ds \quad (i = 1, \ldots, N),$$
$$x_{i,n}(t) = \varphi_i(t) + \int_0^1 K_i(t, s) f_i(s, x_{1,n-1}, \ldots, x_{m,n-1})\, ds$$
$$(i = N+1, \ldots, m). \tag{32.11}$$

If the norm in E is again defined by (32.7), the Lipschitz operators L_{S_n} and L_{T-S_n} corresponding to the operators $S_n = \{S_{ni}\}$ and $T - S_n = \{(T_i - S_{ni})\}$ may be estimated as follows:

$$L_{S_n} = \{L_{S_{ni}}\},$$

$$L_{S_{ni}}x = \begin{cases} \int_0^1 \sum_{j=1}^m |K_{in}(t, s)|\, C_{ij}(s)\, x_j(s)\, ds & (i = 1, \ldots, N), \\ 0 & (i = N+1, \ldots, m); \end{cases}$$

$$L_{T-S_n} = \{L_{(T_i - S_{ni})}\},$$

$$L_{(T_i - S_{ni})}x = \begin{cases} \int_0^1 \sum_{j=1}^m |K_i(t, s) - K_{in}(t, s)|\, C_{ij}(s)\, x_j(s)\, ds & (i = 1, \ldots, N), \\ \int_0^1 \sum_{j=1}^m |K_i(t, s)|\, C_{ij}(s)\, x_j(s)\, ds & (i = N+1, \ldots, m). \end{cases}$$

We now consider system (32.1) under different assumptions. Let the functions $\varphi_i(t)$ be square-integrable on $[0, 1]$, the functions $K_i(t, s)$ square-integrable on the domain $t, s \in [0, 1]$, and the functions $f_i(t, x_1, \ldots, x_m)$ continuous and satisfying condition (32.4); moreover, we assume that

$$\int_0^1 |\varphi_i(t)|^2 dt \leqslant \varphi_i^2 \leqslant \Phi^2; \quad \int_0^1\int_0^1 |K_i(t, s)|^2 dt ds \leqslant K_i^2 \leqslant K^2;$$
$$\sum_{i=1}^m \sum_{j=1}^m \int_0^1 C_{ij}^2(s)\, ds \leqslant C^2. \tag{32.12}$$

We may now treat (32.1) as a nonlinear operator equation in an m-normed space E (see §§22–26), i.e., the space whose elements are sequences of functions $x = \{x_i(t)\}$ which are square-integrable on $[0, 1]$.

We define the norm in E by

$$\| x \| = \sup_i \| x_i(t) \|_{L_2}, \tag{32.13}$$

where $\|x_i(t)\|_{L_2}$ denotes the L_2-norm:

$$\| x_i(t) \|_{L_2} = \sqrt{\int_0^1 |x_i(t)|^2 dt}. \tag{32.14}$$

This is the general case considered in Chapter II, with $B_i = L_2$.

We define the operators P^i as orthogonal projections, each taking an element $x_i \in L_2$ into the subspace spanned by the first $k^{(i)}$ elements of a suitable orthonormal basis $\{e_j^{(i)}(t)\}$, i.e.,

$$P_i x_i = \sum_{j=1}^{k^{(i)}} \alpha_j^{(i)} e_j^{(i)}(t), \qquad \text{where} \qquad \alpha_j^{(i)} = \int_0^1 x_i(t)\, e_j^{(i)}(t)\, dt.$$

The operators $P_N T$, $Q_N T$, ΠT, $Q_\Pi T$, RT and $Q_R T$ are given by the following expressions:

$$P_N T x = \begin{cases} \varphi_i(t) + \int_0^1 K_i(t, s) f_i(s, x_1, \ldots, x_m)\, ds & (i = 1, \ldots, N), \\ 0 \quad (i = N+1, \ldots, m); \end{cases} \tag{32.15}$$

$$Q_N T x = \begin{cases} 0 \quad (i = 1, \ldots, N), \\ \varphi_i(t) + \int_0^1 K_i(t, s) f_i(s, x_1, \ldots, x_m)\, ds \ (i = N+1, \ldots, m); \end{cases} \tag{32.16}$$

$$\Pi T x = \bar{\varphi}_i(t) + \int_0^1 \bar{K}_i(t, s) f(s, x_1, \ldots, x_m)\, ds \quad (i = 1, \ldots, m), \tag{32.17}$$

$$\bar{\varphi}_i(t) = P_i \varphi_i = \sum_{j=1}^{k^{(i)}} \varphi_j^{(i)} e_j^{(i)}(t), \quad \varphi_j^{(i)} = \int_0^1 \varphi_i(t)\, e_j^{(i)}(t)\, dt,$$

$$\bar{K}_i(t, s) = \sum_{j=1}^{k^{(i)}} K_j^{(i)}(s)\, e_j^{(i)}(t), \quad K_j^{(i)}(s) = \int_0^1 K_i(t, s)\, e_j^{(i)}(t)\, dt;$$

$$Q_\Pi T x = \varphi_i(t) - \bar{\varphi}_i(t) + \int_0^1 [K_i(t, s) - \bar{K}_i(t, s)] \times f_i(s, x_1, \ldots, x_m)\, ds \quad (i = 1, \ldots, m); \tag{32.18}$$

$$RTx = \begin{cases} \overline{\varphi}_i(t) + \int_0^1 \overline{K}_i(t, s) f_i(s, x_1, \ldots, x_m)\, ds \quad (i = 1, \ldots, N), \\ 0 \quad (i = N+1, \ldots, m); \end{cases} \tag{32.19}$$

$$Q_R Tx = \begin{cases} \varphi_i(t) - \overline{\varphi}_i(t) + \int_0^1 [K_i(t, s) - \overline{K}_i(t, s)] f_i(s, x_1, \ldots, x_m)\, ds \\ \qquad (i = 1, \ldots, N), \\ \varphi_i(t) + \int_0^1 K_i(t, s) f_i(s, x_1, \ldots, x_m)\, ds \\ \qquad (i = N+1, \ldots, m). \end{cases} \tag{32.20}$$

Solving system (32.1) by the projection method (25.9) with the projection R, we obtain the following system of equations for the approximate solution $y = \{y_i(t)\}$:

$$\begin{aligned} y_i(t) &= \overline{\varphi}_i(t) + \int_0^1 \overline{K}_i(t, s) f_i(s, y_1, \ldots, y_m)\, ds \quad (i = 1, \ldots, N), \\ y_i(t) &= 0 \quad (i = N+1, \ldots, m). \end{aligned} \tag{32.21}$$

If we are using Sokolov's method (17.1) with $k = 1$, the successive approximations $x_n(t) = \{x_{i,n}(t)\}$ are determined from the system

$$\begin{aligned} x_{i,n}(t) &= \varphi_i(t) + \int_0^1 [K_i(t, s) - \overline{K}_i(t, s)] f_i(s, x_{1,n-1}, \ldots, x_{m,n-1})\, ds \\ &\quad + \int_0^1 \overline{K}_i(t, s) f_i(s, x_{1,n}, \ldots, x_{m,n})\, ds \quad (i = 1, \ldots, N), \\ x_{i,n}(t) &= \varphi_i(t) + \int_0^1 K_i(t, s) f_i(s, x_{1,n-1}, \ldots, x_{m,n-1})\, ds \\ &\qquad (i = N+1, \ldots, m). \end{aligned} \tag{32.22}$$

The more general algorithm (7.13), with projection R and $k = 1$, is written in this case as

$$\begin{aligned} x_{i,n}(t) &= \varphi_i(t) + \int_0^1 [K_i(t, s) - \overline{K}_i(t, s)] f_i(s, P_1' x_{1,n}, \ldots, \\ &\qquad \ldots, P_N' x_{N,n}, Q_{N+1}' x_{N+1,n-1}, \ldots, Q_m' x_{m,n-1})\, ds \\ &\quad + \int_0^1 \overline{K}_i(t, s) f_i(s, x_{1,n}, \ldots, x_{m,n})\, ds \quad (i = 1, \ldots, N), \\ x_{i,n}(t) &= \varphi_i(t) + \int_0^1 K_i(t, s) f_i(s, P_1' x_{1,n}, \ldots \\ &\qquad \ldots, P_N' x_{N,n}, Q_{N+1}' x_{N+1,n-1}, \ldots, Q_m' x_{m,n-1})\, ds, \\ &\qquad (i = N+1, \ldots, m), \end{aligned} \tag{32.23}$$

where $P'_i x_i$ is the projection of x_i on the subspace spanned by the first $l^{(i)}$ elements of a basis $\{g_j^{(i)}\}$, and $Q'_i x_i = x_i - P'_i x_i$.

An estimate for the Lipschitz operator q_T of T may be derived as follows:

$$\|Tx - Ty\| = \sup_i \|T_i x - T_i y\|_{L_2}$$

$$\leqslant \sup_i \sqrt{\int_0^1 \left[\int_0^1 |K_i(t,s)|\,|f_i(s, x_1, \ldots, x_m) - f_i(s, y_1, \ldots, y_m)|ds\right]^2 dt}$$

$$\leqslant \sup_i \sqrt{\int_0^1 \left[\int_0^1 |K_i(t,s)| \sum_{j=1}^m C_{ij}(s)\,|x_j(s) - y_j(s)|\,ds\right]^2 dt}$$

$$\leqslant \sup_i \sqrt{\int_0^1 \left[\sum_{j=1}^m \sqrt{\int_0^1 |K_i(t,s)|^2 C_{ij}^2(s)ds} \cdot \sqrt{\int_0^1 |x_j(s) - y_j(s)|^2 ds}\right]^2 dt}$$

$$\leqslant \sup_i \sqrt{\int_0^1 \left[\sum_{j=1}^m \sqrt{\int_0^1 |K_i(t,s)|^2 C_{ij}^2(s)\,ds} \sup_j \|x_j(s) - y_j(s)\|_{L_2}\right]^2 dt}$$

$$= \sup_i \sqrt{\int_0^1 \left[\sum_{j=1}^m \sqrt{\int_0^1 |K_i(t,s)|^2 C_{ij}^2(s)\,ds}\right]^2 dt}\;\|x - y\|. \tag{32.24}$$

Consequently,

$$q_T \leqslant \bar{q}_T = \sup_i \sqrt{\int_0^1 \left[\sum_{j=1}^m \sqrt{\int_0^1 |K_i(t,s)|^2 C_{ij}^2(s)\,ds}\right]^2 dt}\,.$$

Similar manipulations yield estimates for $q_{P_N T}$, $q_{Q_{P_N} T}$, $q_{\Pi T}$, $q_{Q_\Pi T}$, q_{RT}, q_{QRT} and $q_{P_N Q_R T}$. Thus:

$$q_{P_N T} \leqslant \bar{q}_{P_N T} = \sup_{i=1,\ldots,N} \sqrt{\int_0^1 \left[\sum_{j=1}^m \sqrt{\int_0^1 |K_i(t,s)|^2 C_{ij}^2(s)\,ds}\right]^2 dt}\,,$$

$$q_{Q_{P_N}T} \leqslant \bar{q}_{Q_{P_N}T} = \sup_{i=N+1,\ldots,m} \sqrt{\int_0^1 \left[\sum_{j=1}^m \sqrt{\int_0^1 |K_i(t,s)|^2 C_{ij}^2(s)\,ds}\right]^2 dt}\,,$$

$$q_{\Pi T} \leqslant \bar{q}_{\Pi T} = \sup_{i=1,\ldots,m} \sqrt{\int_0^1 \left[\sum_{j=1}^m \sqrt{\int_0^1 |\bar{K}_i(t,s)|^2 C_{ij}^2(s)\,ds}\right]^2 dt}\,,$$

$$q_{Q_\Pi T} \leqslant \bar{q}_{Q_\Pi T} = \sup_{i=1,\ldots,m} \sqrt{\int_0^1 \left[\sum_{j=1}^m \sqrt{\int_0^1 |K_i(t,s) - \bar{K}_i(t,s)|^2 C_{ij}^2(s)ds}\right]^2 dt}\,,$$

$$q_{RT} \leqslant \bar{q}_{RT} = \sup_{i=1,\ldots,N} \sqrt{\int_0^1 \left[\sum_{j=1}^m \sqrt{\int_0^1 |\bar{K}_i(t,s)|^2 C_{ij}^2(s)\,ds}\right]^2 dt}\,,$$

$$q_{Q_RT} \leqslant \bar{q}_{Q_RT} = \sup_{i=1,\ldots,m} \begin{cases} \sqrt{\int_0^1 \left[\sum_{j=1}^m \sqrt{\int_0^1 |K_i(t,s) - \overline{K}_i(t,s)|^2 C_{ij}^2(s)\,ds}\right]^2 dt}. \\ \qquad (i = 1, \ldots, N), \\ \sqrt{\int_0^1 \left[\sum_{j=1}^m \sqrt{\int_0^1 |K_i(t,s)|^2 C_{ij}^2(s)\,ds}\right]^2 dt} \\ \qquad (i = N+1, \ldots, m), \end{cases}$$

$$q_{P_NQ_RT} \leqslant \bar{q}_{P_NQ_RT} = \sup_{i=1,\ldots,N} \sqrt{\int_0^1 \left[\sum_{j=1}^m \sqrt{\int_0^1 |K_i(t,s) - \overline{K}_i(t,s)|^2 C_{ij}^2(s)\,ds}\right]^2 dt}.$$

Since L_2 is a Hilbert space and P_i is by assumption an orthogonal projection, we can make use of the convergence criterion (26.12).

Suppose that in some domain $\|x_i - \varphi_i\|_{L_2} \leqslant M_i$ we have estimates

$$|f_i(s, x_1, \ldots, x_m)| \leqslant F^{(i)}(M_1, \ldots, M_m). \tag{32.25}$$

Then, using the Cauchy-Bunjakovskiĭ-Schwarz inequality, we can proceed as follows:

$$F_{RT}^{(i)}(M_1, \ldots, M_m) = L_R^{(i)} F^{(i)}(M_1, \ldots, M_m) \quad (i = 1, \ldots, N),$$

$$F_{P_NQ_RT}^{(i)}(M_1, \ldots, M_m) = L_{P_NQ_R}^{(i)} F^{(i)}(M_1, \ldots, M_m) \quad (i = 1, \ldots, N),$$

$$F_{Q_{P_N}T}^{(i)}(M_1, \ldots, M_m) = L_{Q_{P_N}}^{(i)} F^{(i)}(M_1, \ldots, M_m) \quad (i = N+1, \ldots, m),$$

where

$$L_R^{(i)} = \sqrt{\int_0^1 \left[\int_0^1 |\overline{K}_i(t,s)|\,ds\right]^2 dt},$$

$$L_{P_NQ_R}^{(i)} = \sqrt{\int_0^1 \left[\int_0^1 |K_i(t,s) - \overline{K}_i(t,s)|\,ds\right]^2 dt},$$

$$L_{Q_{P_N}}^{(i)} = \sqrt{\int_0^1 \left[\int_0^1 |K_i(t,s)|\,ds\right]^2 dt}.$$

Consequently, we can stipulate that the constants M_i form a solution of the system of inequalities (see (26.14))

$$\begin{gathered} ([L_R^{(i)}]^2 + [L_{P_NQ_R}^{(i)}]^2)\,[F^{(i)}(M_1, \ldots, M_m)]^2 \leqslant M_i^2 \\ (i = 1, \ldots, N), \\ L_{Q_{P_N}}^{(i)} F^{(i)}(M_1, \ldots, M_m) \leqslant M_i \quad (i = N+1, \ldots, m). \end{gathered} \tag{32.26}$$

Note that under the conditions (32.12) system (32.1) may also be treated as a nonlinear operator equation in the Hilbert space H with norm

$$\|x\| = \sqrt{\sum_{i=1}^m \|x_i\|_{L_2^{(i)}}^2}, \tag{32.27}$$

where

$$\| x_i \|^2_{L_2^{(i)}} = \int_0^1 \rho_i(t) \, | x_i(t) |^2 dt \quad (\rho_i(t) > 0). \tag{32.28}$$

It is clear that in this case the projections P_N, Q_{P_N}, Π, Q_Π, R and Q_R defined above are orthogonal, and so the estimates developed in Chapter II for Hilbert spaces are applicable.

Thus, using standard inequalities, we find

$$\| Tx - Ty \|^2 \leqslant \sum_{i=1}^m \int_0^1 \rho_i(t) \left[\int_0^1 | K_i(t, s) | \, | f_i(s, x_1, \ldots, x_m) - f_i(s, y_1, \ldots, y_m) | \, ds \right]^2 dt$$

$$\leqslant \sum_{i=1}^m \int_0^1 \rho_i(t) \left[\int_0^1 | K_i(t, s) | \sum_{j=1}^m C_{ij}(s) \, | x_j(s) - y_j(s) | \, ds \right]^2 dt$$

$$\leqslant \sum_{i=1}^m \int_0^1 \rho_i(t) \left[\sum_{j=1}^m \sqrt{\int_0^1 | K_i(t, s) |^2 \frac{C_{ij}^2(s)}{\rho_j(s)} ds} \times \sqrt{\int_0^1 \rho_j(s) \, | x_j(s) - y_j(s) |^2 ds} \right]^2 dt$$

$$\leqslant \sum_{i=1}^m \int_0^1 \rho_i(t) \left[\sum_{j=1}^m \int_0^1 | K_i(t, s) |^2 \frac{C_{ij}^2(s)}{\rho_j(s)} ds \sum_{j=1}^m \int_0^1 \rho_j(s) \, | x_j(s) - y_j(s) |^2 ds \right] dt$$

$$= \sum_{i=1}^m \int_0^1 \sum_{j=1}^m \int_0^1 \rho_i(t) \, | K_i(t, s) |^2 \frac{C_{ij}^2(s)}{\rho_j(s)} ds dt \, \| x - y \|^2. \tag{32.29}$$

Consequently,

$$q_T \leqslant \sqrt{\sum_{i=1}^m \int_0^1 \sum_{j=1}^m \int_0^1 \rho_i(t) \, | K_i(t, s) |^2 \frac{C_{ij}^2(s)}{\rho_j(s)} ds dt} \, .$$

Similar procedures yield estimates for the Lipschitz constants of the other operators.

If $\| K_i \|^{(i)}_{L_2}$ are the norms of the corresponding operators

$$K_i x_i = \int_0^1 K_i(t, s) \, x_i(s) \, ds,$$

we have the estimate

$$q_T^2 \leqslant \sup_s \sum_{i=1}^m \sum_{j=1}^m \| K_i \|_{L_2}^2 \rho_i(s) \frac{C_{ij}^2(s)}{\rho_j(s)}.$$

Similar estimates are valid for the other operators.

EXAMPLE 12. The system of nonlinear integral equations

$$\begin{aligned} x_1(t) &= \frac{13}{20} t + \frac{21}{160} + \int_0^1 \left(t - \frac{1}{2} s\right) \left[x_1^2(s) + \frac{1}{5} x_2(s) \right] ds, \\ x_2(t) &= \frac{3}{20} t^2 + \frac{1}{2} \int_0^1 t^2 s x_1^3(s)\, ds \end{aligned} \tag{32.30}$$

has a solution $x_1^*(t) = t$, $x_2^*(t) = t^2/4$.

Let us solve system (32.30) by the projection method, with the projection R, setting $N = 1$ and defining the P_i by

$$P_1 x_1 = 3t \int_0^1 s x(s)\, ds, \quad P_2 = 0.$$

Then (25.9) becomes

$$y_1(t) = \frac{271}{320} t + \int_0^1 t \left(1 - \frac{3}{4} s\right) y_1^2(s)\, ds, \qquad y_2(t) = 0. \tag{32.31}$$

Solving (32.31) in the domain $|y_1| < 1$, $|y_2| < 1$, we find

$$y_1(t) \approx 0.98973t, \qquad y_2(t) = 0.$$

In this example we have $Tx = \{T_i x\}$, $i = 1, 2$, where

$$T_1 x = \frac{13}{20} t + \frac{21}{160} + \int_0^1 \left(t - \frac{1}{2} s\right) \left[x_1^2(s) + \frac{1}{5} x_2(s) \right] ds;$$

$$T_2 x = \frac{3}{20} t^2 + \frac{1}{2} \int_0^1 t^2 s x_1^3(s)\, ds;$$

$$P_N T x = \begin{cases} \dfrac{13}{20} t + \dfrac{21}{160} + \displaystyle\int_0^1 \left(t - \frac{1}{2} s\right) \left[x_1^2(s) + \frac{1}{5} x_2(s) \right] ds & (i = 1); \\ 0 \quad (i = 2), \end{cases}$$

$$Q_{P_N} T x = \begin{cases} 0 \quad (i = 1), \\ \dfrac{3}{20} t^2 + \dfrac{1}{2} \displaystyle\int_0^1 t^2 s x_1^3(s)\, ds \quad (i = 2); \end{cases}$$

$$RTx = \begin{cases} \dfrac{271}{320} t + \displaystyle\int_0^1 t \left(1 - \frac{3}{4} s\right) \left[x_1^2(s) + \frac{1}{5} x_2(s) \right] ds & (i = 1), \\ 0 \quad (i = 2); \end{cases}$$

$$Q_R T x = \begin{cases} \dfrac{21}{160} - \dfrac{63}{320} t + \dfrac{1}{2} \displaystyle\int_0^1 \left(\frac{3}{2} t - 1\right) s \left[x_1^2(s) + \frac{1}{5} x_2(s) \right] ds \quad (i = 1), \\ \dfrac{3}{20} t^2 + \dfrac{1}{2} \displaystyle\int_0^1 t^2 s x_1^3(s)\, ds \quad (i = 2); \end{cases}$$

$$P_N Q_R T x = \begin{cases} \dfrac{21}{160} - \dfrac{63}{320} t + \dfrac{1}{2} \int\limits_0^1 \left(\dfrac{3}{2} t - 1\right) s \left[x_1^2(s) + \dfrac{1}{5} x_2(s)\right] ds \\ \qquad\qquad\qquad\qquad (i = 1), \\ 0 \quad (i = 2). \end{cases}$$

If system (32.30) is treated as a nonlinear operator equation in an m-normed space, with B_i the space L_2, we find the following errors:

$$\|x_1^* - y_1\|_{L_2} \approx 0.0103 \cdot \frac{1}{\sqrt{3}}; \qquad \|x_2^* - y_2\|_{L_2} = \frac{1}{4\sqrt{5}};$$

$$\|x^* - y\| = \frac{1}{4\sqrt{5}} \approx 0.1118.$$

Since clearly

$$Q_{P_N} T y = \{0; 0.2470 t^2\}, \quad P_N Q_R T y = \{0.0088 - 0.0132 t; 0\}$$

and

$$q_{P_N T} \leqslant \frac{\sqrt{6}}{6}\left(2M_1 + \frac{1}{5}\right), \quad q_{Q_{P_N} T} \leqslant \frac{\sqrt{15}}{10} M_1^2 \ (\text{for } |x_1| \leqslant M),$$

it follows that for $M_1 = 1$, say,

$$q_{P_N T} \leqslant \frac{11\sqrt{6}}{30} \approx 0.898; \qquad q_{Q_{P_N} T} \leqslant \frac{\sqrt{15}}{10} \approx 0.387$$

and by inequality (25.10) we obtain

$$\|x^* - y\| \leqslant (1 - q_{Q_{P_N} T})^{-1} \|Q_{P_N} T y\| \approx 1.631 \cdot 0.247 \cdot \frac{1}{\sqrt{3}} \approx 0.23.$$

EXAMPLE 13. A solution of the system of nonlinear integral equations

$$\begin{aligned} x_1(t) &= 5 - \frac{3}{4} t - \int\limits_0^1 (12 + ts)[s^2 x_1(s) + x_3(s)]\, ds, \\ x_2(t) &= t^2 + \int\limits_0^1 (24t^2 + s^2)[s x_2(s) - x_3(s)]\, ds, \\ x_3(t) &= \frac{899}{900} t^3 + \frac{1}{10} \int\limits_0^1 t^3 s^3 x_1(s) x_2(s) x_3(s)\, ds \end{aligned} \tag{32.32}$$

in the domain $|x_i(t)| < 1$ $(i = 1, 2, 3)$ is $x_1(t) = 1 - t$, $x_2(t) = t^2$, $x_3(t) = t^3$.

We define a projection R by $Rx = \{P_1 x_1, P_2 x_2, 0\}$, where

$$P_1 x_1 = \int\limits_0^1 x_1(s)\, ds, \qquad P_2 x_2 = 5 \int\limits_0^1 t^2 s^2 x_2(s)\, ds.$$

The algorithm

$$x_n = R T x_n + Q_R T x_{n-1} \qquad (Q_R = I - R), \tag{32.33}$$

considered in §26 may be written for system (32.32) as

$$x_{1,n}(t) = 5 - \frac{3}{4}t - \int_0^1 \left(12 + \frac{1}{2}s\right)[s^2 x_{1,n}(s) + x_{3,n}(s)\, ds$$
$$- \int_0^1 \left(t - \frac{1}{2}\right) s\,[s^2 x_{1,n-1}(s) + x_{3,n-1}(s)]\, ds, \tag{32.34}$$

$$x_{2,n}(t) = t^2 + \int_0^1 t^2\left(24 + \frac{5}{3}s^2\right)[sx_{2,n}(s) - x_{3,n}(s)]\, ds$$
$$+ \int_0^1 \left(1 - \frac{5}{3}t^2\right) s^2\,[sx_{2,n-1}(s) - x_{3,n-1}(s)]\, ds,$$

$$x_{3,n}(t) = \frac{899}{900}t^3 + \frac{1}{10}\int_0^1 t^3 s^3 x_{1,n-1}(s)\, x_{2,n-1}(s)\, x_{3,n-1}(s)\, ds.$$

This system yields

$$x_{1,n}(t) = \frac{38\,056}{46\,125} - \frac{3}{4}t - \frac{62}{1025}\int_0^1 s^3 x_{1,n-1}(s)\, x_{2,n-1}(s)\, x_{3,n-1}(s)$$
$$+ \int_0^1 \left(\frac{144}{205} - t\right) s\,[s^2 x_{1,n-1}(s) + x_{3,n-1}(s)]\, ds;$$

$$x_{2,n}(t) = \frac{85\,387}{85\,500}t^2 + \frac{113}{950}\int_0^1 t^2 s^3 x_{1,n-1}(s)\, x_{2,n-1}(s)\, x_{3,n-1}(s)$$
$$+ \int_0^1 \left(1 - \frac{387}{190}t^2\right) s^2\,[sx_{2,n-1}(s) - x_{3,n-1}(s)]\, ds;$$

$$x_{3,n}(t) = \frac{899}{900}t^3 + \frac{1}{10}\int_0^1 t^3 s^3 x_{1,n-1}(s)\, x_{2,n-1}(s)\, x_{3,n-1}(s).$$

Using the convergence criteria of the classical method of successive approximations, one readily shows that if the initial approximations $x_{i,0}(t)$ $(i = 1, 2, 3)$ are selected sufficiently near the respective functions $x_i(t)$ $(i = 1, 2, 3)$ in absolute value, then the sequence $\{x_n(t)\}$ $(x_n(t) = \{x_{i,n}(t)\})$ converges to a solution $x(t) = \{x_i(t)\}$ of system (32.32). At the same time, the usual method of successive approximations applied directly to system (32.32) is divergent.

If we set $x_{1,0}(t) = ½$, $x_{2,0}(t) = 1/3$ and $x_{3,0}(t) = 1/4$, the result is

$$x_{1,1}(t) = 1.000042 - t; \quad x_{2,1}(t) = 0.999954t^2; \quad x_{3,1}(t) = 0.999931t^3.$$

§33. Forced oscillations of finite amplitude

Forced oscillations of a pendulum in the absence of dissipative forces are described [215] by the nonlinear differential equation

$$v'' + \alpha^2 \sin v = F(t), \tag{33.1}$$

where α^2 is a parameter and $F(t)$ a periodic applied force.

Suppose that the applied force $F(t)$ is an odd function with period 2ω, say, $F(t) = \beta \sin(\pi t/\omega)$. We then have the following fundamental problem: Does equation (33.1) possess a solution of the same type–an odd 2ω-periodic function? This problem is equivalent to solution of equation (33.1) with boundary conditions

$$v(0) = v(1) = 0 \tag{33.2}$$

(in the case $\omega = 1$).

Since the Green's function for the problem

$$x''(t) + \lambda r(t) = 0, \quad x(0) = x(1) = 0$$

is

$$G(t, s) = \begin{cases} s(1-t) & (0 \leqslant s \leqslant t), \\ t(1-s) & (t \leqslant s \leqslant 1), \end{cases}$$

it follows that problem (33.1)–(33.2) is equivalent to the integral equation

$$v(t) = -\int_0^1 G(t, s)\,[F(s) - \alpha^2 \sin v(s)]\, ds.$$

Performing the substitution

$$\int_0^1 G(t, s) F(s)\, ds = g(t), \quad v(t) + g(t) = x(t), \tag{33.3}$$

we obtain a Hammerstein equation

$$x(t) = \alpha^2 \int_0^1 G(t, s) \sin [x(s) - g(s)]\, ds. \tag{33.4}$$

In this case,

$$Tx = \alpha^2 \int_0^1 G(t, s) \sin [x(s) - g(s)]\, ds.$$

Let us treat (33.4) as a nonlinear operator equation in L_2. The kernel $G(t, s)$ is symmetric, its eigenvalues are the numbers $1/n^2\pi^2$ ($n = 1, 2, \ldots$), and the function $f(t, x) = \alpha^2 \sin(x - g)$ satisfies the condition

$$\max \left| \frac{\partial f(t, x)}{\partial x} \right| = \max |\alpha^2 \cos (x - g)| = \alpha^2, \tag{33.5}$$

and this implies the following estimate for the Lipschitz constant of the operator T:

$$q_T \leqslant \frac{\alpha^2}{\pi^2}. \tag{33.6}$$

Consequently, if $\alpha < \pi$ the classical successive approximations converge to the unique solution of (33.4) in L_2, at least as rapidly as a geometric progression with quotient α^2/π^2.

It follows from the results of § 17 that the algorithms (17.1) and (17.3) with $k = 1$ are also convergent, for any choice of the orthogonal projection P.

Let us estimate the rate of convergence of these algorithms for one special choice of the projection: projection onto the subspace spanned by the first k elements of an orthonormal system of eigenfunctions of $G(t, s)$. Since $G(t, s)$ may be written as a bilinear series

$$G(t, s) = 2\sum_{n=1}^{\infty} \frac{\sin n\pi t \sin n\pi s}{(n\pi)^2},$$

it follows that

$$PTx = 2\alpha^2 \int_0^1 \sum_{n=1}^{k} \frac{\sin n\pi t \sin n\pi s}{(n\pi)^2} \sin [x(s) - g(s)]\, ds, \qquad (33.7)$$

$$QTx = 2\alpha^2 \int_0^1 \sum_{n=k+1}^{\infty} \frac{\sin n\pi t \sin n\pi s}{(n\pi)^2} \sin [x(s) - g(s)]\, ds. \qquad (33.8)$$

The norm of a symmetric linear operator is equal to the maximum of the absolute value of its eigenvalues, and so, using (33.5), we find estimates for q_{PT} and q_{QT}:

$$q_{PT} \leqslant \frac{\alpha^2}{\pi^2}; \quad q_{QT} \leqslant \frac{\alpha^2}{(k+1)^2\pi^2}.$$

Consequently, the rate of convergence of algorithm (17.1) for $k = 1$, which in this case may be written

$$x_n(t) = \alpha^2 \int_0^1 [G(t, s) - \bar{G}(t, s)] \sin [x_{n-1}(s) - g(s)]\, ds$$
$$+ \alpha^2 \int_0^1 \bar{G}(t, s) \sin [x_n(s) - g(s)]\, ds, \qquad (33.9)$$

is at least that of a geometric progression with quotient

$$\varepsilon_k = \frac{\alpha^2}{(k+1)^2\pi^2 \sqrt{1 - \frac{\alpha^4}{\pi^4}}} = \frac{\alpha^2}{(k+1)^2 \sqrt{\pi^4 - \alpha^4}}.$$

Here

$$\bar{G}(t, s) = 2\sum_{n=1}^{k} \frac{\sin n\pi t \sin n\pi s}{(n\pi)^2}.$$

To estimate the successive approximations, we find the following expressions:

a) For the ordinary method of successive approximations,

$$\eta_n(t) = \alpha^2 \int_0^1 G(t, s) \sin [\eta_{n-1}(s) - g(s)]\, ds \qquad (33.10)$$

we have

$$|x^*(t) - \eta_n(t)| \leqslant \bar{G}(t) \frac{\alpha^2\pi^2}{\pi^2 - \alpha^2} \left(\frac{\alpha^2}{\pi^2}\right)^{n-p} \|\delta_p\|_{L_2} \quad (1 \leqslant p \leqslant n). \qquad (33.11)$$

where

$$\delta_p = \eta_p - \eta_{p-1};\ \overline{G}(t) = \sqrt{\int_0^1 |G(t, s)|^2 ds}\,.$$

b) For the algorithm (33.9),

$$|x^*(t) - x_n(t)| \tag{33.12}$$
$$\leqslant \alpha^2 [\overline{Q_k G(t)} + \varepsilon_k \overline{P_k G(t)}] \frac{\varepsilon_k^{n-p}}{1-\varepsilon_k} \|\overline{\delta}_p\|_{L_2} \quad (1 \leqslant p \leqslant n),$$

where

$$\overline{Q_k G(t)} = \sqrt{\int_0^1 \left| 2 \sum_{n=k+1}^{\infty} \frac{\sin n\pi t \sin n\pi s}{(n\pi)^2} \right|^2 ds}\,,$$

$$\overline{P_k G(t)} = \sqrt{\int_0^1 \left| 2 \sum_{n=1}^{k} \frac{\sin n\pi t \sin n\pi s}{(n\pi)^2} \right|^2 ds},\ \overline{\delta}_p = x_p - x_{p-1}\,.$$

Similar procedures yield an error estimate for the projection method, and one can also deduce convergence criteria and error estimates for the more general iterative projection methods considered in Chapters I and II.

§34. Boundary-value problem of the theory of thermal explosion

In the theory of thermal explosion we encounter the boundary-value problem

$$\Delta T = \frac{q}{k} e^{-\frac{\alpha}{T}} = 0 \tag{34.1}$$

in a domain D, with

$$T = T_0 > 0 \tag{34.2}$$

on the boundary Γ of D.

Here T is the temperature, q the thermal effect, k the thermal conductivity of the gaseous mixture, and α the activation energy.

The solution of this problem yields a temperature distribution for each prescribed temperature of the container's walls. At a certain temperature T_0, known as the ignition temperature, this distribution becomes impossible.

In the one-dimensional case, conditions under which problem (34.1), (34.2) is solvable were found by Frank-Kameneckiĭ. For an infinite container with plane-parallel walls, the equation may be integrated in general form for any temperature distribution. The equation in this case is

$$\frac{d^2T}{dx^2} = -kv(T), \tag{34.3}$$

and its integral is

$$x = \int \frac{dT}{\sqrt{-2\int kv(\tau)\,d\tau}}. \tag{34.4}$$

An estimate for this condition valid for any number of dimensions was derived by Anisimov in [4].

If we denote

$$\tau(\bar{r}) = \frac{1}{\alpha} T(\bar{r});\ \tau_0(\bar{r}) = \frac{1}{\alpha} T_0(\bar{r});\ \lambda = \frac{q}{\alpha k},$$

then problem (34.1), (34.2) is equivalent to solution of the nonlinear integral equation

$$\tau(\bar{r}) = f(\bar{r}) + \lambda \int_D G(\bar{r}, \bar{r}') e^{-\frac{1}{\tau(\bar{r}')}} d\bar{r}', \tag{34.5}$$

where

$$f(\bar{r}) = \int_\Gamma \tau_0(\bar{r}') \frac{\partial G(\bar{r}, \bar{r}')}{\partial n} d\bar{r}',$$

$\partial/\partial n$ denotes differentiation with respect to the inner normal to Γ, and $G(\bar{r}, \bar{r}')$ is the Green's function of the Laplacian: $\Delta G = -\delta(\bar{r} - \bar{r}')$ in D and $G = 0$ on Γ. Anisimov in [4] used the classical method of successive approximations to solve equation (34.5):

$$\begin{gathered} \tau_n(\bar{r}) = f(\bar{r}) + \lambda \int_D G(\bar{r}, \bar{r}') e^{-\frac{1}{\tau_{n-1}(\bar{r}')}} d\bar{r}', \\ \tau_0(\bar{r}) = 0 \end{gathered} \tag{34.6}$$

and obtained a convergence criterion

$$\lambda < \tau_m^2 e^{\frac{1}{\tau_m}} A_G^{-1}, \tag{34.7}$$

where $A_G = \max \int_D G(\bar{r}, \bar{r}')\, d\bar{r}'$, and τ_m is the maximum temperature in D. (It is known that the case of interest in chemical kinetics is $\tau_m \ll 1$.)

Using the results of [63], we can deduce a sharper estimate:

$$\lambda < \tau_m^2 e^{\frac{1}{\tau_m}} \lambda_G^{-1}, \tag{34.8}$$

where λ_G is the eigenvalue of $G(\bar{r}, \bar{r}')$ of maximum absolute value. The successive approximations $\tau_n(\bar{r})$ then converge to the required solution of (34.5) at least as rapidly as a geometric progression with quotient

$$|\lambda| \lambda_G \tau_m^{-2} e^{-\frac{1}{\tau_m}}.$$

To determine an approximate bound $\bar{\tau}_m$ for τ_m, we can use the inequality

$$\max f(\bar{r}) + |\lambda| A_G e^{-\frac{1}{\bar{\tau}_m}} \leqslant \bar{\tau}_m. \tag{34.9}$$

If we solve (34.5) by the iterative process

$$\tau_n(\bar{r}) = f(\bar{r}) + \lambda \int_D [G(\bar{r}, \bar{r}') - \bar{G}(\bar{r}, \bar{r}')]\, e^{-\frac{1}{\tau_{n-1}(\bar{r}')}}\, d\bar{r}' + \lambda \int_D \bar{G}(\bar{r}, \bar{r}')\, e^{-\frac{1}{\tau_n(\bar{r}')}}\, d\bar{r}', \tag{34.10}$$

where $\bar{G}(\bar{r}, \bar{r}')$ is a symmetric function approximating $G(\bar{r}, \bar{r}')$, then, defining the bound $\bar{\tau}_m$ by the inequality

$$\max f(\bar{r}) + |\lambda| (A_{G-\bar{G}} + A_{\bar{G}})\, e^{-\frac{1}{\bar{\tau}_m}} \leqslant \bar{\tau}_m, \tag{34.11}$$

where

$$A_{G-\bar{G}} = \max \int_D |G(\bar{r}, \bar{r}') - \bar{G}(r, \bar{r}')|\, d\bar{r}', \quad A_{\bar{G}} = \max \int_D |\bar{G}(\bar{r}, \bar{r}')|\, d\bar{r}',$$

we have the following convergence criterion for the process (34.10):

$$|\lambda| (\lambda_{G-\bar{G}} + \lambda_{\bar{G}})\, \bar{\tau}_m^{-2} e^{-\frac{1}{\bar{\tau}_m}} < 1, \tag{34.12}$$

where $\lambda_{\bar{G}}$ and $\lambda_{G-\bar{G}}$ are the eigenvalues of $\bar{G}(\bar{r}, \bar{r}')$ and $G(\bar{r}, \bar{r}') - \bar{G}(\bar{r}, \bar{r}')$, respectively, of largest absolute value.

The rate of convergence is bounded below by that of a geometric progression with quotient

(34.13)
$$\varepsilon = \frac{|\lambda|\, \lambda_{G-\bar{G}}\, \bar{\tau}_m^{-2} e^{-\frac{1}{\bar{\tau}_m}}}{1 - |\lambda|\, \lambda_{\bar{G}}\, \bar{\tau}_m^{-2} e^{-\frac{1}{\bar{\tau}_m}}}. \tag{34.13}$$

It also follows from § 17 that if $\bar{\tau}_m$ is defined by (34.11), then condition (34.8), with $\bar{\tau}_m$ in place of τ_m, may serve as a convergence criterion for the process (34.10).

EXAMPLE. Consider the domain $0 \leqslant x \leqslant a;\ 0 \leqslant y \leqslant b$. The Green's function for the Laplacian in this domain is

$$G(x, y; \xi, \eta) = \frac{4}{\pi^2 ab} \sum_{m,n} \frac{\sin \frac{m\pi x}{a} \sin \frac{n\pi y}{b} \sin \frac{m\pi \xi}{a} \sin \frac{n\pi \eta}{b}}{\left(\frac{m}{a}\right)^2 + \left(\frac{n}{b}\right)^2}$$

or

$$G(x, y; \xi, \eta) = \frac{2}{\pi} \sum_{n=1}^{\infty} \frac{\sin \frac{n\pi x}{a} \sin \frac{n\pi \xi}{b}}{n \operatorname{sh} \frac{n\pi b}{a}} \times \begin{cases} \sinh \frac{n\pi \xi}{a} \sinh \left[\frac{n\pi}{a} (b - \eta)\right], & \eta > y; \\ \sinh \frac{n\pi \eta}{a} \sinh \left[\frac{n\pi}{a} (b - y)\right], & \eta < y. \end{cases}$$

Since

$$A_G = \frac{a^2}{8}\left[1 - \frac{32}{\pi^3}\cdot\frac{1}{\cosh\frac{\pi b}{2a}}\right],$$

we may write condition (34.7) as

$$\lambda < \frac{8\tau_m^2 e^{\frac{1}{\tau_m}}}{a^2\left(1 - \frac{32}{\pi^3}\cdot\frac{1}{\cosh\frac{\pi b}{2a}}\right)}.$$

The sharper condition (34.8) is here

$$\lambda < \tau_m^2 e^{\frac{1}{\tau_m}}\pi^2\left(\frac{1}{a^2} + \frac{1}{b^2}\right).$$

If $b = \infty$, we have, respectively,

$$\lambda < \frac{8}{a^2}\tau_m^2 e^{\frac{1}{\tau_m}} \quad\text{and}\quad \lambda < \frac{\pi^2}{a^2}\tau_m^2 e^{\frac{1}{\tau_m}}.$$

If we define $\overline{G}(x, y; \xi, \eta)$ in (34.10) by

$$\overline{G}(x, y; \xi, \eta) = \frac{4}{\pi^2 ab}\sum_{m, n}^{m', n'}\frac{\sin\frac{m\pi x}{a}\sin\frac{n\pi y}{b}\sin\frac{m\pi\xi}{a}\sin\frac{n\pi\eta}{b}}{\left(\frac{m}{a}\right)^2 + \left(\frac{n}{b}\right)^2}$$

and note that the eigenvalues of maximum absolute value for G, $\overline{G}$ and $G - \overline{G}$ are, respectively,

$$\frac{1}{\pi^2\left(\frac{1}{a^2} + \frac{1}{b^2}\right)},\quad \frac{1}{\pi^2\left(\frac{1}{a^2} + \frac{1}{b^2}\right)} \quad\text{and}\quad \frac{1}{\pi^2\left[\left(\frac{m'+1}{a}\right)^2 + \left(\frac{n'+1}{b}\right)^2\right]},$$

we see that the rate of convergence of the successive approximations (34.10) is at least that of a geometric progression with quotient

$$\varepsilon = \frac{\bar{\tau}_m^{-2} e^{-\frac{1}{\bar{\tau}_m}}}{\pi^2\left[\left(\frac{m'+1}{a}\right)^2 + \left(\frac{n'+1}{b}\right)^2\right]\sqrt{1 - \frac{\bar{\tau}_m^{-4} e^{-\frac{2}{\bar{\tau}_m}}}{\pi^4\left(\frac{1}{a^2} + \frac{1}{b^2}\right)^2}}},$$

or, in the case $b = \infty$,

$$\varepsilon = \frac{\tau_m^{-2} e^{-\frac{1}{\tau_m}}}{\pi^2\left(\frac{m'+1}{a}\right)^2\sqrt{1 - \frac{a^4}{\pi^4}\tau_m^{-4} e^{-\frac{2}{\tau_m}}}}.$$

§35. Systems of nonlinear integral equations of the theory of shallow shells of revolution

As shown, e.g., in [16], investigation of the bending of shallow shells of revolution under the action of an arbitrary normal load $p(x)$ is equivalent, for small deflections, to solution of a system of nonlinear integral equations

$$\theta(x) = -\int_{\alpha}^{1} k_1(x, \xi)\,[f(\xi) - k^2(\xi)\,s(\xi) + \theta(\xi)\,s(\xi)]\,d\xi, \tag{35.1}$$

$$s(x) = -\int_{\alpha}^{1} k_2(x, \xi)\left[k^2(\xi)\,\theta(\xi) - \frac{\theta^2(\xi)}{2}\right] d\xi,$$

where

$$f(\xi) = -\lambda \int_{\alpha}^{\xi} p(\xi)\,\xi d\xi,$$

$$\rho = \frac{r}{a};\ \alpha = \frac{b}{a};\ \theta = \frac{1}{h}\sqrt{12(1-\mu^2)}\,\frac{dw}{d\xi},\quad s = \frac{12(1-\mu^2)\,a^2}{Eh^3}\,\xi N_r,$$

$$k_2(\xi) = \frac{1}{h}\sqrt{12(1-\mu^2)}\,\frac{dz(\xi)}{d\xi},\quad \lambda = \frac{a^4\,[12(1-\mu^2)]^{\frac{3}{2}}}{Eh^4}.$$

The notation used here is as follows:

r is the distance from an arbitrary point of the neutral surface to its axis of rotation.

b and a are the inner and outer radii, respectively, of the bases of the shell.

$z(\xi)$ is the equation of a generator of the neutral surface.

h, w, N_r are respectively the thickness, deflection and radial membrane strain.

E, μ denote the elastic modulus and Poisson ratio of the shell material.

$k_1(x, \xi)$ and $k_2(x, \xi)$ are the Green's functions of the first-order Bessel equation satisfying the same boundary conditions as the unknown functions $\theta(x)$ and $s(x)$:

$$k_1(x, \xi) = \sum_{i=1}^{\infty} \frac{z_1(\lambda_i x)\,z_1(\lambda_i \xi)}{\lambda_i^2 \int_{\alpha}^{1} x z_1^2(\lambda_i x)\,dx}, \tag{35.2}$$

$$k_2(x, \xi) = \sum_{i=1}^{\infty} \frac{z_1(\gamma_i x)\,z_1(\gamma_i \xi)}{\gamma_i^2 \int_{\alpha}^{1} x z_1^2(\gamma_i x)\,dx}, \tag{35.3}$$

where

$$z_1(\lambda_i x) = c_1 I_1(\lambda_i x) + c_2 N_1(\lambda_i x),\quad z_1(\gamma_i x) = c_1' I_1(\gamma_i x) + c_2' N_1(\gamma_i x),$$

I_1 and N_1 are first-order Bessel functions of the first and second kind, respectively; c_1, c_2, c_1', c_2', γ_i and λ_i depend on the end conditions.

System (35.1) may be treated as a nonlinear operator equation, say in a coordinate space E whose elements are pairs of functions from suitable function spaces. We shall assume that both functions are in L_2. If the norm of an element $u(x) = \{\theta(x), s(x)\}$ in E is defined by

$$\|u\|^2 = \|\theta(x)\|^2_{L_2} + \|s(x)\|^2_{L_2},$$

$$\|\theta(x)\|^2_{L_2} = \int_\alpha^1 x\,|\theta(x)|^2 dx, \quad \|s(x)\|^2 = \int_\alpha^1 x\,|s(x)|^2 dx,$$

then E is a Hilbert space.

Using the contraction mapping principle, one can show that, if certain additional restrictions are imposed on the functions $f(x)$ and $k^2(x)$, system (35.1) will have a solution of sufficiently small norm.

If moreover the Lipschitz constant of the operator in question is less than unity, this solution is unique in an appropriate ball. It may be determined by successive approximations if the initial approximations $\theta_0(x)$ and $s_0(x)$ are functions of sufficiently small norm.

Alternatively, one can employ algorithm (17.1) or other algorithms studied in Chapters I and II, defining the projection P as $P = \{P_i\}$ $(i = 1, 2)$, where P_1 is the orthogonal projection onto the subspace spanned by the first k elements of a set of orthonormal eigenfunctions of $k_1(x, \xi)$, and P_2 the orthogonal projection onto the span of the first k' elements of a set of orthonormal eigenfunctions of $k_2(x, \xi)$.

According to this algorithm, the successive approximations $\theta_n(x)$ and $s_n(x)$ are defined by a nonlinear system

$$\theta_n(x) = -\int_\alpha^1 [k_1(x, \xi) - k_1^{(k)}(x, \xi)]\,[f(\xi) - k^2(\xi)\,s_{n-1}(\xi) + \theta_{n-1}(\xi)\,s_{n-1}(\xi)]\,d\xi$$

$$-\int_\alpha^1 k_1^{(k)}(x, \xi)\,[f(\xi) - k^2(\xi)\,S_n(\xi) + \theta_n(\xi)\,s_n(\xi)]\,d\xi, \tag{35.4}$$

$$s_n(x) = -\int_\alpha^1 [k_2(x, \xi) - k_2^{(k')}(x, \xi)]\left[k^2(\xi)\,\theta_{n-1}(\xi) - \frac{\theta^2_{n-1}(\xi)}{2}\right]d\xi -$$

$$-\int_\alpha^1 k_2^{(k')}(x, \xi)\left[k^2(\xi)\,\theta_n(\xi) - \frac{\theta^2_n(\xi)}{2}\right]d\xi.$$

If system (35.1) is solved by the projection method with the above projection P, the approximations $\theta^{(k)}(x)$ and $s^{(k')}(x)$ are defined by the system

$$\theta^{(k)}(x) = -\int_\alpha^1 k_1^{(k)}(x, \xi)\,[f(\xi) - k^2(\xi)\,s^{(k')}(\xi) + \theta^{(k)}(\xi)\,s^{(k')}(\xi)]\,d\xi,$$

$$s^{(k')}(x) = -\int_\alpha^1 k_2^{(k')}(x, \xi)\left[k^2(\xi)\,\theta^{(k)}(\xi) - \frac{\theta^{(k)2}(\xi)}{2}\right]d\xi,$$

where

$$k_1^{(k)}(x,\xi)=\sum_{i=1}^{k}\frac{z_1(\lambda_i x)\,z_1(\lambda_i\xi)}{\lambda_i^2\int_\alpha^1 xz_1^2(\lambda_i x)\,dx},\quad k_2^{(k')}(x,\xi)=\sum_{i=1}^{k'}\frac{z_1(\gamma_i x)\,z_1(\gamma_i\xi)}{\gamma_i^2\int_\alpha^1 xz_1^2(\gamma_i x)\,dx}.$$

Since P is an orthogonal projection, we may use the results of §§16 and 17.

BIBLIOGRAPHY

1. R. F. Albrecht and G. Karrer, *Fixpunktsätze in uniformen Räumen*, Math. Z. 74 (1960), 387–391. MR 23 #A1359.

2. M. Altman, *On the approximate solution of non-linear functional equations*, Bull. Acad. Polon. Sci. Cl. III 5 (1957), 457–460. MR 19, 984.

3. ———, *Concerning approximate solutions of non-linear functional equations*, Bull. Acad. Polon. Sci. Cl. III 5 (1957), 461–465. MR 19, 984.

4. I. S. Anisimov, *Stationary temperature distribution in the presence of a chemical reaction*, Dokl. Akad. Nauk BSSR 5 (1961), 380–382. (Russian)

5. P. M. Anselone and R. H. Moore, *Approximate solutions of integral and operator equations*, J. Math. Anal. Appl. 9 (1964), 268–277. MR 32 #1920.

6. A. B. Bakušinskiĭ, *A method for the numerical solution of integral equations*, Comput. Methods and Programming (Comput. Center Moscow Univ. Collect. Works III), Izdat. Moskov. Univ., Moscow, 1965, pp. 536–543. (Russian) MR 33 #1985.

7. A. N. Baluev, *On a method of Čaplygin*, Vestnik Leningrad. Univ. 11 (1956), no. 13, 27–42. (Russian) MR 18, 321.

8. Stefan Banach, *Sur les opérations dans les ensembles abstraits et leur application aux équations intégrales*, Fund. Math. 3 (1922), 133–181.

9. K. B. Barataliev, *The approximate solution of integro-differential equations with deviating argument*, Proc. Third Siberian Conf. Math. Mech. (1964), Izdat. Tomsk. Univ., Tomsk, 1964, pp. 248–250. (Russian) MR 37 #5661.

10. ———, *Application of Ju. D. Sokolov's method to the solution of integral equations with deviating argument*, Trudy Frunze Politehn. Inst. Vyp. 21 (1965), 59–69. (Russian) RŽMat. 1966 #11Б283.

11. ———, *Approximate solution of two-dimensional linear integral equations with deviating argument by Ju. D. Sokolov's method*, Materials Thirteenth Sci. Conf. Professors and Instructors Phys.-Math. Faculty (Math. Section), "Mektep", Frunze, 1965, pp. 16–18. (Russian) RŽMat. 1966 #9Б350.

12. ———, *The approximate solution of integro-differential equations with deviating argument*, Studies in Integro-Differential Equations in Kirghizia, no. 3, "Ilim", Frunze, 1965, pp. 69–83. (Russian) MR 37 #5662.

13. ———, *Approximate solution of some problems for integro-differential equations with retarded argument*, Candidate's Dissertation, Akad. Nauk Kirgiz. SSR, Frunze, 1965. (Russian)

14. B. A. Bel'tjukov, *Construction of rapidly converging iterative algorithms for the solution of integral equations*, Sibirsk. Mat. Ž. 6 (1965), 1415–1419. (Russian) MR 33 #4625.

15. I. S. Berezin and N. P. Židkov, *Computing methods*. Vol. 2, Fizmatgiz, Moscow, 1959; English transl., Addison-Wesley, Reading, Mass.; Pergamon Press, New York, 1966. MR 30 #4372; 31 #1756.

16. A. A. Berezovskiĭ, *Nonlinear integral equations of sloping shells of revolution,* Inž. Ž. 1 (1961), no. 4, 107–114. (Russian)

17. G. D. Birkhoff, *Lattice theory*, 3rd ed., Amer. Math. Soc. Colloq. Publ., vol. 25, Amer. Math. Soc., Providence, R.I., 1967. MR 37 #2638.

18. L. P. Bogdanova, *On the approximate solution of a class of nonlinear integral equations with constant limits,* Proc. Sci. Conf. Engrs., Aspirants and Junior Assistants Inst. Math. Sci. Ukrain. SSR, Izdat. Inst. Mat. Akad. Nauk Ukrain. SSR, Kiev, 1963, pp. 63–72. (Russian)

19. P. S. Bondarenko, *Study of numerical algorithms for the approximate integration of differential equations by the method of finite differences,* Vidav. Kiiv. Univ., Kiev, 1962. (Ukrainian)

20. F. E. Browder, *Fixed point theorems for nonlinear semi-contractive mappings in Banach spaces,* Arch. Rational Mech. Anal. 21 (1966), 259–269. MR 34 #641.

21. F. E. Browder and W. V. Petryshyn, *The solution by iteration of linear functional equations in Banach spaces*, Bull. Amer. Math. Soc. 72 (1966), 566–570. MR 32 #8155a.

22. ———, *The solution by iteration of nonlinear functional equations in Banach spaces*, Bull. Amer. Math. Soc. 72 (1966), 571–575. MR 32 #8155b.

23. R. Caccioppoli, *Sugli elementi uniti delle transformazioni funzionali: un'osservazione sui problemi di valori ai limiti,* Atti Reale Accad. Naz. Lincei (6) 13 (1931), 498–502.

24. P. Ĭ. Čalenko, *Estimation of the error in the method of strips for the approximate solution of integral equations,* Vīsnik Kiïv. Univ. 1962, no. 5, Ser. Mat. Meh. vyp. 2, 70–78. (Ukrainian) MR 34 #5330.

25. ———, *Solution of integral equations by the method of strips*, Candidate's Dissertation, Kiev. State Univ., Kiev, 1963. (Russian)

26. ———, *Finding the eigenfunctions of Fredholm integral equations of the second kind*, Vīsnik Kiïv. Univ. 1964, no. 6, 95–101. (Ukrainian) MR 32 #8076.

27. ———, *On the error of the method of strips in the application of numerical integration formulas,* Vyčisl. Mat. (Kiev) Vyp. 1 (1965), 79–89. (Russian) MR 34 #8641.

28. S. A. Čaplygin, *A new method for approximate integration of differential equations*, GITTL, Moscow, 1950. (Russian)

29. E. A. Černyšenko, *Investigation of convergence and establishment of an estimate of the error of the method of averaging in a complete normed space*, Ukrain. Mat. Ž. 6 (1954), 305–313. (Russian) MR 17, 665.

30. ———, *The method of averaging applied to the determination of eigenvalues of an operator equation,* Dopovīdī Akad. Nauk Ukraïn. RSR 1955, 217–221. (Ukrainian) MR 17, 665.

31. ———, *On some methods of approximate solution of operator equations*, Candidate's Dissertation, Inst. Mat. Akad. Nauk Ukrain. SSR, Kiev, 1955. (Russian) RŽMat. 1956 #4667.

32. ———, *On a variant of the method of averaging,* Dopovīdī Akad. Nauk Ukraïn. RSR 1956, 10–12. (Ukrainian) MR 17, 901.

33. ———, *On a method of approximate solution of Cauchy's problem for ordinary differential equations*, Ukrain. Mat. Ž. 10 (1958), no. 1, 89–100. (Russian) MR 20 #4919.

34. ———, *On the approximate solution of a boundary-value problem for the heat equation,* Naučn. Soobšč. Dnepropetrovsk. Inž.-Stroit. Inst., Vyp. 41 (1959). (Russian) RŽMat. 1960 #11726.

35. L. Cesari, *Functional analysis and Galerkin's method,* Michigan Math. J. 11 (1964), 385–414. MR 30 #4047.

36. L. Collatz, *Einige Anwendungen funktionalanalytischer Methoden in der praktischen Analysis,* Z. Angew. Math. Physik 4 (1953), 327–357. MR 15, 559.

37. ———, ***Funktionalanalysis und numerische Mathematik***, Die Grundlehren der math. Wissenschaften, Band 120, Springer-Verlag, Berlin, 1964; English transl., Academic Press, New York, 1966. MR 29 #2931; 34 #4961.

38. ———, *Einige abstrakte Begriffe in der numerischen Mathematik* (*Anwendungen der Halbordnung*), Computing (Arch Elektron. Rechnen) 1 (1966), 233–255. MR 34 #3746.

39. ———, *Numerisch Behandlung von Differentialgleichungen*, Die Grundlehren der math. Wissenschaften, Band 60, Springer-Verlag, Berlin, 1951. MR 13, 285.

40. R. Courant and D. Hilbert, *Methoden der mathematischen Physik,* Vol. 1, 2nd ed., Vol. 2, Springer, Berlin, 1931, 1937.

41. R. Courant, K. O. Friedrichs and H. Lewy, *Über die partiellen Differenzengleichungen der matematischen Physik,* Math. Ann. 100 (1928), 32–74; English transl., IBM J. Res. Develop. 11 (1967), 215–234. MR 35 #4621.

42. Romulus Cristescu, *Sur la méthode des approximations successives pour des équations aux opérateurs*, Acad. R. P. Romîne Fil. Cluj Stud. Cerc. Mat. 12 (1961), 41–44.

43. B. P. Demidovič, I. A. Maron and E. Z. Šuvalova, *Numerical methods of analysis,* Fizmatgiz, Moscow, 1962. (Russian) MR 39 #1071.

44. V. Ju. Dīdik, *An approximate method for solving operator equations of a special type,* Dopovīdī Akad. Nauk Ukraïn. RSR Ser. A 1967, 490–492. (Ukrainian) MR 35 #2508.

45. C. L. Dolph, *Nonlinear integral equations of the Hammerstein type*, Trans. Amer. Math. Soc. 66 (1949), 289–307. MR 11, 367.

46. C. L. Dolph and G. J. Minty, *On nonlinear integral equations of the Hammerstein type,* Nonlinear Integral Equations (Proc. Advanced Seminar Conducted by Math. Research Center, U. S. Army, Univ. of Wisconsin, Madison, Wis., 1963), Univ. of Wisconsin Press, Madison, Wis., 1964, pp. 99–154. MR 28 #4322.

47. L. N. Dovbyš (Gagen-Torn) and S. G. Mihlin, *On the solvability of nonlinear Ritz systems,* Dokl. Akad. Nauk SSSR 138 (1961), 258–260. (Russian) MR 24 #B910.

48. M. Edelstein, *An extension of Banach's contraction principle,* Proc. Amer. Math. Soc. 12 (1961), 7–10. MR 22 #11375.

49. Hans Ehrmann, *Ein abstrakter Satz zur Konvergenzerzeugung und Konvergenzverbesserung für Iterationsverfahren bei nichtlinearen Gleichungen,* Z. Angew. Math. Mech. 37 (1957), 252–254.

50. ———, *Iterationsverfahren mit veränderlichen Operatoren,* Arch Rational Mech. Anal. 4 (1959), erratum, ibid. 6 (1960), 440. MR 22 #311; errata, 22, p. 2545.

51. D. K. Faddeev and V. N. Faddeeva, *Computational methods of linear algebra*, 2nd ed., Fizmatgiz, Moscow, 1963; English transl., Freeman, San Francisco, Calif., 1963. MR 28 #1742; #4659.

52. G. E. Forsythe and W. R. Wasow, *Difference methods of solution of partial differential equations*, Wiley, New York, 1960. MR 23 #B3156.

53. T. Frey, *Über ein neues Iterationsverfahren zur Lösung von Integral- und Intergrodifferentialgleichungen,* Sympos. Numerical Treatment of Ordinary Differential Equations, Integral and Integro-Differential Equations (Rome, 1960), Birkhäuser, Basel, 1960, pp. 384-387. MR 23 #B2606.

54. ———, *Study of the Picard and Carathéodory iteration methods*, Tájékoztató Magyár Tud. Akad. Számítastechn. Közp. 1966, no. 10, 27–95. (Hungarian) RŽMat. 1967 #2Б560.

55. V. M. Fridman, *New methods of solving a linear operator equation*, Dokl. Akad. Nauk SSSR 128 (1959), 482–484. (Russian) MR 22 #1066.

56. M. M. Gal′, *A sufficient condition for the convergence of Ju. D. Sokolov's method for integral equations with deviating argument*, Abstracts of Papers of Drogobych State Ped. Inst. Eight Periodic Sci. Conf. (1966), Drogobič. Derž. Ped. Inst., Drogobych, 1966, pp. 59–63. (Ukrainian) MR 36 #7359.

57. ———, *Approximate solution of integral equations with retarded argument by the method of Ju. D. Sokolov*, Ukrain. Mat. Ž. 18 (1966), no. 6, 102–107. (Russian) MR 34 #5331.

58. ———, *Justification of the application of the method of averaging of functional corrections to determination of the spectral density of an error of an extremal pulse system with modulation*, Ukrain. Mat. Ž. 19 (1967), no. 3, 95–103. (Russian) MR 39 #7839.

59. B. G. Galerkin, *Rods and plates. Series in some problems of elastic equivalence of rods and plates*, Vestnik Inž. A1915, 897–908; reprinted in *Collected works*. Vol. 1, Izdat. Akad. Nauk SSSR, Moscow, 1952, pp. 168–195. (Russian)

60. M. K. Gavurin and L. V. Kantorovič, *Approximational and numerical methods*, Forty Years of Mathematics in the USSR: 1917–1957. Vol. 1: Survey articles, Fizmatgiz, Moscow, 1959, pp. 809–855. (Russian)

61. Michael Golomb, *Zur Theorie der nichtlinearen Integralgleichungen, Integralgleichungssysteme und allgemeinen Funktional-gleichungen*, Math. Z. 39 (1934/35), 45–75.

62. Ju. I. Gribanov, *On the projection method for solving linear equations in Banach spaces*, Izv. Vysš. Učebn. Zaved. Matematika 1965, no. 1 (44), 48–55. (Russian) MR 30 #4137.

63. A. Hammerstein, *Nichtlinear Integralgleichungen nebst Anwendungen*, Acta Math. 54 (1930), 117–176.

64. R. I. Kačurovskiĭ, *On some fixed-point principles*, Moskov. Oblast. Ped. Inst. Učen. Zap. 96 (1960), 215–219. (Russian)

65. A. F. Kalaĭda, *Research on and application of mean-value theorems*, Candidate's Dissertation, Inst. Kibernet. Akad. Nauk Ukrain. SSR, Kiev, 1965. (Russian)

66. O. F. Kalaĭda and V. Ju. Sereda, *A general method for solving linear integral equations of the second kind*, Dopovidī Akad. Nauk Ukraïn. SSR Ser. A 1967, 492–495. (Ukrainian) MR 35 #2509.

67. L. V. Kantorovič, *Functional analysis and applied mathematics*, Uspehi Mat. Nauk 3 (1948), no. 6 (28), 89–185; English transl., Nat. Bur. Standards Rep. no. 1509, U. S. Dept. of Commerce, Nat. Bur. Standards, Washington, D. C., 1952. MR 10, 380; 14, 766.

68. ———, *On some further applications of the Newton approximation method*, Vestnik Leningrad. Univ. Ser. Mat. Meh. Astr. 12 (1957), no. 7, 68–103. (Russian) MR 19, 883.

69. L. V. Kantorovič and G. P. Akilov, *Functional analysis in normed spaces*, Fizmatgiz, Moscow, 1959; English transl., Internat. Ser. Monographs Pure and Appl. Math., vol. 46, Macmillan, New York, 1964. MR 22 #9837; 35 #4699.

70. L. V. Kantorovič and V. I. Krylov, *Approximate methods of higher analysis*, 5th ed., Fizmatgiz, Moscow, 1962; English transl. of 3rd ed., Interscience, New York; Noordhoff, Groningen, 1958. MR 21 #5268; 27 #4338.

71. L. V. Kantorovič, B. Z. Vulih and A. G Pinsker, *Functional analysis in semi-ordered spaces*, GITTL, Moscow, 1950. (Russian) MR 12, 340.

72. M. V. Keldyš, *On Galerkin's method of solution of boundary problems*, Izv. Akad. Nauk SSSR Ser. Mat. 6 (1942), 309–330. (Russian) MR 5, 7.

73. I. I. Kolodner, *Equations of Hammerstein type in Hilbert spaces*, J. Math. Mech. 13 (1964), 701–750. MR 30 #1415.

74. J. Kolomý, *Some existence theorems for nonlinear problems*, Comment. Math. Univ. Carolinae 7 (1966), 207–217. MR 34 #6580.

75. Z. Kowalski, *An iterative method of solving differential equations*, Ann. Polon. Math. 12 (1962/63), 213–230. MR 26 #2013.

76. M. A. Krasnosel'skiĭ, *Convergence of Galerkin's method for nonlinear equations*, Dokl. Akad. Nauk SSSR 73 (1950), 1121–1124. (Russian) MR 12, 187.

77. ———, *Two remarks on the method of successive approximations*, Uspehi Mat. Nauk 10 (1955), no. 1 (63), 123–127. (Russian) MR 16, 833.

78. ———, *Topological methods in the theory of nonlinear integral equations*, GITTL, Moscow, 1956; English transl., Macmillan, New York, 1964. MR 20 #3464; 28 #2414.

79. M. A. Krasnosel'skiĭ and S. G. Kreĭn, *An iteration process with minimal residuals*, Mat. Sb. 31 (73) (1952), 315–334. (Russian) MR 14, 692.

80. M. A. Krasnosel'skiĭ and Ja. B. Rutickiĭ, *Certain approximate methods for solving nonlinear operator equations*, Proc. Fourth All-Union Math. Congr. (Leningrad, 1961), vol. II, "Nauka" Leningrad, 1964, pp. 562–571. (Russian) MR 36 #4399.

81. M. A. Krasnosel'skiĭ, P. P. Zabreĭko, E. I. Pustyl'nik and P. E. Sobolevskiĭ, *Integral operators in spaces of summable functions*, "Nauka", Moscow, 1966. (Russian) MR 34 #6568.

82. M. Kravčuk, *Sur la résolution des équations linéaires différentielles et intégrales par la méthode de moments.* I, II, Izdat. Akad. Nauk Ukrain. SSR, Kiev, 1932, 1936. (Ukrainian)

83. L. E. Krivošein, *On the approximate solution of some integro-differential equations*, Učen. Zap. Fiz.-Mat. Fak. Kirgiz. Univ. Vyp. 4 (1957), part 2, 39–68. (Russian) RŽMat. 1958 #9263.

84. ———, *Approximate solution of some problems for linear integro-differential equations*, Candidate's Dissertation, Central Asian State Univ., Frunze, 1958. (Russian) RŽMat. 1959 #7504.

85. ———, *Approximate methods for solution of ordinary linear integro-differential equations*, Izdat. Akad. Nauk Kirgiz. SSR, Frunze, 1962. (Russian) RŽMat. 1962 #11Б160.

86. L. E. Krivošein and K. B. Barataliev, *On the approximate solution of nonlinear two-dimensional integro-differential equations with deviating argument*, Materials Thirteenth Sci. Conf. Professors and Instructors Phys.-Math. Faculty (Math. Section), "Mektep", Frunze, 1965, pp. 50–53. (Russian) RŽMat. 1966 #9Б357.

87. M. M. Krylov, *Sur différentes généralisations de la méthode de W. Ritz et de la méthode de moindres carrés pour l'intégration approchée des équations de la physique mathématique*, Ukraïn. Akad. Nauk Trudi Fiz.-Mat. Viddilu 3 (1926/27), no. 2. (Ukrainian)

88. ———, *Approximate solution of the basic problems of mathematical physics*, Izdat. Akad. Nauk Ukrain. SSR, Kiev, 1931. (Russian)

89. N. M. Krylov and N. N. Bologoljubov, *Sur le calcul des racines de la transcendante de Fredholm les plus voisines d'un nombre donné par les méthodes des moindres carrés et de

l'algorithme variationnel, Izv. Akad. Nauk SSSR Otdel. Fiz.-Mat. Nauk (7) 1929, 471–488.

90. N. S. Kurpel', *On some approximate methods for solution of nonlinear equations in special Banach spaces,* Proc. Sci. Conf. Engrs., Aspirants and Junior Research Assistants Inst. Math. Acad. Sci. Ukrain. SSR, Izdat. Inst. Mat. Akad. Nauk Ukrain. SSR, Kiev, 1963, pp. 57–62. (Russian)

91. ———, *On the approximate solution of non-linear operator equations by the method of Ju. D. Sokolov,* Ukrain. Mat. Ž. 15 (1963), 309–314. (Russian) MR 29 #1546.

92. ———, *Estimate of the error of the projection method and Ju. D. Sokolov's method for nonlinear equations in a coordinate space,* Dopovīdī Akad. Nauk Ukrain. RSR 1963, no. 9, 1135–1139. (Ukrainian)

93. ———, *An approximate method of solving linear operator equations in Hilbert space,* Dopovīdī Akad. Nauk Ukraïn. RSR 1963, 1275–1279. (Ukrainian) MR 29 #2652.

94. ———, *Sufficient conditions for the convergence of the method of Ju. D. Sokolov for approximate solution of non-linear integral equations of Hammerstein type,* Approximate Methods of Solving Differential Equations, Izdat. Akad. Nauk Ukrain. SSR, Kiev, 1963, pp. 47–53. (Russian) MR 28 #5803.

95. ———, *On some approximate methods of solution of non-linear operator equations,* Candidate's Dissertation, Inst. Mat. Akad. Nauk Ukrain. SSR, Kiev, 1963. (Russian)

96. ———, *Some approximate methods of solving non-linear equations in a coordinate Banach space,* Ukrain. Mat. Ž. 16 (1964), 115–120. (Russian) MR 28 #4324.

97. ———, *Conditions for convergence and error estimates for a general iterative method of solution of linear operator equations,* First Republ. Math. Conf. of Young Researchers, part I, Akad. Nauk Ukrain. SSR Inst. Mat., Kiev, 1965, pp. 418–427. (Russian) MR 33 #4731.

98. ———, *Existence and uniqueness of solutions of a class of nonlinear operator equations,* First Republ. Math. Conf. of Young Researchers, part I, Akad. Nauk Ukrain. ISSR Inst. Mat., Kiev, 1965, pp. 428–434. (Russian) MR 33 #4732.

99. ———, *On a generalization of the method of averaging functional corrections,* Dopovīdī Akad. Nauk Ukraïn. RSR 1965, 1005–1009. (Ukrainian) MR 33 #6864.

100. ———, *The convergence and error estimates of certain general iterative methods of solution of operator equations,* Dopovīdī Akad. Nauk Ukraïn. RSR 1965, 1423–1427. (Ukrainian) MR 33 #6808.

101. ———, *Some general approximate methods for solving operator equations,* Internat. Congress Math., Moscow, 1966, Abstracts, Section 14, p. 39. (Russian)

102. ———, *Convergence and error estimates for an iterative projection method of solving operator equations,* Proc. Second Sci. Conf. Young Ukrainian Mathematicians, "Naukova Dumka", Kiev, 1966, pp. 361–365. (Ukrainian)

103. ———, *On an iterative projection method of solution of operator equations,* Dopovīdī Akad. Nauk Ukraïn. RSR Ser. A 1967, 218–222. (Ukrainian) MR 35 #3500.

104. M. Kwapisz, *On the iterative method of solving differential equations with retardations,* Prace Mat. 9 (1965), 57–68. MR 30 #5021.

105. ———, *On a certain iterative method of approximations and qualitative problems for functional-differential and differential equations in Banach space,* Zeszyty Nauk Politech. Gdánsk. 79, Mat. 4 (1955), 3–73. (Polish)

106. ———, *On the approximate solutions of an abstract equation,* Ann. Polon. Math. 19 (1967), 47–60. MR 35 #1207.

107. O. A. Ladyženskaja, *The method of finite differences in the theory of partial differential equations*, Uspehi Mat. Nauk 12 (1957), no. 5 (77), 123–148; English transl., Amer. Math. Soc. Transl. (2) 20 (1962), 77–104. MR 20 #3395; 25 #342.

108. A. Langenbah, *On an application of the method of least squares to nonlinear equations*, Dokl. Akad. Nauk SSSR 143 (1962), 31–34 = Soviet Math. Dokl. 3 (1962), 330–334. MR 26 #2878.

109. V. I. Lebedev, *On the KP-method of improving the convergence of iterations in solving the kinetic equation*, Ž. Vyčisl. Mat. i Mat. Fiz. 6 (1966), no. 4, suppl., 154–176. (Russian) MR 36 #1127.

110. J. Leray and J. Schauder, *Topologie et équations fonctionelles*, Ann. Sci. École Norm. Sup. (3) 51 (1934), 45–78.

111. L. A. Ljusternik and V. I. Sobolev, *Elements of functional analysis*, 2nd rev. ed., "Nauka", Moscow, 1965; English transl. of 1st ed., Ungar, New York, 1961. MR 25 #5361; 35 #698.

112. A. Ju. Lučka, *A sufficient condition for the convergence of the procedure for averaging functional corrections*, Dokl. Akad. Nauk SSSR 122 (1958), 179–182. (Russian) MR 20 #4922.

113. ———, *Approximate solution of Fredholm integral equations by the method of averaged functional corrections*, Ukrain. Mat. Ž. 12 (1960), 32–45. (Russian) MR 28 #4316.

114. ———, *Approximate solution of linear operator equations in a Banach space by Ju. D. Sokolov's method*, Ukrain. Mat. Ž. 13 (1961), no. 1, 39–52. (Russian) MR 28 #5307.

115. ———, *The approximate solution of infinite systems of algebraic equations by Ju. D. Sokolov's method*, Dopovīdī Akad. Nauk Ukraïn. RSR 1961, 146–149. (Ukrainian) MR 23 #B566.

116. ———, *Approximate solution of linear operator equations in a Banach space by Ju. D. Sokolov's method*, Dopovīdī Akad. Nauk Ukraïn. RSR 1961, 424–428. (Ukrainian) MR 26 #4181.

117. ———, *On the theory and applications of the method of averaging functional corrections*, Candidate's Dissertation, Akad. Nauk Ukrain. SSR, Kiev, 1961. (Russian)

118. ———, *Approximate solution of infinite systems of linear integral equations by Ju. D. Sokolov's method*, Dopovīdī Akad. Nauk Ukraïn. RSR 1962, 1149–1153. (Ukrainian) MR 26 #1720.

119. ———, *Approximate solution of infinite systems of linear differential equations by the method of Ju. D. Sokolov*, Dopovīdī Akad. Nauk Ukraïn. RSR 1963, 563–567. (Ukrainian) MR 34 #5298.

120. ———, *Theory and application of the method of averaging functional corrections*, Izdat. Akad. Nauk Ukrain. SSR, Kiev, 1963; English transl., Academic Press, New York, 1965. MR 30 #4394; 32 #3298.

121. ———, *On an application of the method of Ju. D. Sokolov to the solution of a Dirichlet problem for the Poisson equation*, Dopovīdī Akad. Nauk Ukraïn. RSR 1965, 426–429. (Ukrainian) MR 32 #6696.

122. ———, *Application of the method of Ju. D. Sokolov to the solution of the exterior Neumann problem for the Poisson equation*, Dopovīdī Akad. Nauk Ukraïn. RSR 1965, 547–550. (Ukrainian) MR 32 #2753.

123. A. Ju. Lučka, *Approximate solution of operator equations by an iteration projection method,* Internat. Congr. Math., Moscow, 1966, Abstracts, Section 14, p. 41. (Russian)

124. A. Ju. Lučka and N. S. Kurpel', *On a non-stationary iteration method for the approximate solution of linear operator equations,* Ukrain. Mat. Ž. 16 (1964), 389–395. (Russian) MR 29 #3885.

125. A. E. Martynjuk, *Certain approximate methods of solution of nonlinear equations with unbounded operators,* Izv. Vysš. Učebn. Zaved. Matematika 1966, no. 6 (55), 85–94. (Russian) MR 34 #6573.

126. S. G. Mihlin, *Direct methods in mathematical physics,* GITTL, Moscow, 1950. (Russian) MR 16, 41.

127. ———, *Variational methods in mathematical physics,* GITTL, Moscow, 1957; English transl., Macmillan, New York, 1964. MR 22 #1881; 30 #2712.

128. ———, *The numerical performance of variational methods,* "Nauka", Moscow, 1966; English transl., Wolters-Noordhoff, Groningen, 1971. MR 34 #3747; 43 #4236.

129. Ju. M. Molokovič, *An approximation method for the solution of linear integral equations,* Izv. Vysš. Učebn. Zaved. Matematika 1959, no. 5 (12), 164–170. (Russian) MR 27 #3098.

130. ———, *The approximate solution of linear integral equations by means of a certain variant of Ju. D. Sokolov's method,* Kazan. Gos. Univ. Učen. Zap. 127 (1967), kn. 1, 139–147. (Russian) MR 41 #1255.

131. B. G. Mosolov, *On an approximation method of solving a non-linear integro-differential equation,* Izv. Akad. Nauk UzSSR Ser. Fiz.-Mat. 1961, no. 2, 41–51. (Russian) MR 31 #5053.

132. ———, *On an approximate method of solving linear operator equations in the metric space* L_2, Izv. Akad. Nauk UzSSR Ser. Fiz.-Mat. Nauk 1961, no. 5, 29–34. (Russian) MR 31 #589.

133. ———, *Approximate solution of operator equations by the method of Ju. D. Sokolov,* Izv. Akad. Nauk UzSSR Ser. Fiz.-Mat. Nauk 1963, no. 5, 26–29. (Russian) MR 29 #3002.

134. ———, *On approximate solution of linear weighted integral equations by Ju. D. Sokolov's method,* Integration Certain Differential Equations Math. Phys., Izdat. "Nauka", Uzbek. SSR, Tashkent, 1964, pp. 168–182. (Russian) MR 34 #980.

135. ———, *Approximate solution of some functional equations by the method of averaging functional corrections,* Candidate's Dissertation, Akad. Nauk Uzbek. SSR, Tashkent, 1965. (Russian)

136. ———, *Approximate solution of operator equations of the first kind by the method of averaging functional corrections,* Voprosy Vyčisl. Mat. i Tehn., vyp. 7, "Nauka", Tashkent 1965, pp. 20–26. (Russian) RŽMat. 1966 #5Б496.

137. ———, *Approximate solution of systems of nonlinear integro-differential equations by the method of averaging functional corrections,* Voprosy Kibernet. i. Vyčisl. Mat., vyp. 2, "Fan", Tashkent, 1966, pp. 66–73. (Russian) RŽMat. 1967 #7Б493.

138. A. D. Myškis and I. Ju. Ègle, *On an estimate of the error in the method of successive approximations,* Mat. Sb. 35 (77) (1954), 491–500. (Russian) MR 16, 862.

139. I. P. Mysovskih, *A method of mechanical quadrature for solving integral equations,* Vestnik Leningrad. Univ. 17 (1962), no. 7, 78–88. (Russian) MR 26 #5758.

140. M. Z. Nashed, *On general iterative methods for the solutions of a class of non-*

linear operator equations, Math. Comp. 19 (1965), 14–24. MR 31 #4143.

141. N. N. Nazarov, *Non-linear integral equations of Hammerstein's type*, Acta Univ. Asiae Mediae = Trudy Sredneaziatsk Gos. Univ. Ser. V-a (1941), fasc. 33, 79 pp. (Russian) MR 3, 150.

142. V. V. Nemyckiĭ, *Théorèmes d'existence et d'unicité des solutions de quelques équations intégrales non-linéaires*, Mat. Sb. 41 (1934), 421–438; Russian transl., ibid., 438–452.

143, ———, *The fixed-point method in analysis*, Uspehi Mat. Nauk 1 (1936), 141–174; English transl., Amer. Math. Soc. Transl. (2) 34 (1963), 1–37.

144. Carl Gottfried Neumann, *Untersuchungen über das logarithmische und Newton'sche Potential*, Teubner, Leipzig, 1877.

145. C. Olech, *A connection between two certain methods of successive approximations in differential equations*, Ann. Polon. Math. 11 (1962), 237–245. MR 25 #2258.

146. G. I. Petrov, *Application of Galerkin's method to the problem of stability of flow of a viscous fluid*, Prikl. Mat. Meh. 4 (1940), 3–12. (Russian)

147. W. V. Petryshyn, *Direct and iterative methods for the solution of linear operator equations in Hilbert space*, Trans. Amer. Math. Soc. 105 (1962), 136–175. MR 26 #3180.

148. ———, *On a general iterative method for the approximate solution of linear operator equations*, Math. Comp. 17 (1963), 1–10. MR 29 #729.

149. ———, *On the extension and the solution of nonlinear operator equations*, Illinois J. Math. 10 (1966), 255–274. MR 34 #8242.

150. M. Picone, *Sull'equazione integrale non lineare di Volterra*, Ann. Mat. Pura Appl. (4) 49 (1960), 1–10. MR 22 #2867.

151. ———, *Sull'equazione integrale non lineare di seconda specie di Fredholm*, Math. Z. 74 (1960), 119–128. MR 22 #9826.

152. S. I. Pohožaev, *Analogue of Schmidt's method for nonlinear equations*, Dokl. Akad. Nauk SSSR 136 (1961), 546–548 = Soviet Math. Dokl. 2 (1961), 103–105. MR 22 #5859.

153. G. N. Položiĭ (Editor), *Mathematical practicum*, Fizmatgiz, Moscow, 1960. (Russian) MR 22 #12683.

154. G. M. Položiĭ and P. Ĭ. Čalenko, *The strip method of solving integral equations*, Dopovidī Akad. Nauk Ukraïn. RSR 1962, 427–431. (Ukrainian) MR 28 #5575.

155. ———, *Solution of integral equations by the band method*, Problems Math. Phys. and Theory of Functions, Izdat. Akad. Nauk Ukrain. SSR, Kiev, 1964, pp. 124–144. (Russian) MR 32 #2859.

156. N. I. Pol'skiĭ, *Some generalizations of B. G. Galerkin's method*, Dokl. Akad. Nauk SSSR 86 (1952), 469–472. (Russian) MR 14, 384.

157. ———, *On the convergence of certain approximate methods of analysis*, Ukrain. Mat. Ž. 7 (1955), 56–70. (Russian) MR 17, 64.

158. ———, *On a general scheme of application of approximation methods*, Dokl. Akad. Nauk SSSR 111 (1956), 1181–1184. (Russian) MR 18, 802.

159. ———, *Projective methods in applied mathematics*, Dokl. Akad. Nauk SSSR 143 (1962), 787–790 = Soviet Math. Dokl. 3 (1962), 488–491. MR 26 #3173.

160. R. D. Richtmyer, *Difference methods for initial-value problems*, Interscience Tracts in Pure and Appl. Math., no. 4, Interscience, New York, 1957. MR 20 #438.

161. F. Riesz and B. Sz.-Nagy, *Leçons d'analyse fonctionnelle,* 2nd ed., Akad. Kiadó, Budapest, 1953; English transl., Ungar, New York, 1955. MR **15**, 132; **17**, 175.

162. Walter Ritz, *Über eine neue Methode zur Lösung gewisser Variationsprobleme der mathematischen Physik,* J. Reine Angew. Math. **135** (1909), 1–61.

163. ———, *Gesammelte Werke – Walther Ritz – Oeuvres,* Gauthier-Villars, Paris, 1911.

164. T. Sabirov and A. R. Esajan, *The question of convergence of the averaging method of functional corrections,* Dokl. Akad. Nauk Tadžik SSR **9** (1966), no. 1, 8–12. (Russian) MR **34** #1893.

165. G. S. Salehov and M. A. Mertvecova, *Convergence of certain iterative processes,* Izv. Kazan. Fil. Akad. Nauk SSSR Ser. Fiz.-Mat. Tech. Nauk **1954**, vyp. 5, 77–108; Chinese transl., Advancement in Math. **3** (1957), 341–374. RŽMat. **1956** #2460; MR **20** #7380.

166. H. Schaefer, *Über die Methode der a priori-Schranken,* Math. Ann. **129** (1955), 415–416. MR **17**, 175.

167. J. Schauder, *Der Fixpunktsatz in Funktionalräumen,* Studia Math. **2** (1930), 171–180.

168. J. W. Schmidt, *Konvergenzuntersuchungen und Fehlerabschätzungen für ein verallgemeinertes Iterationsverfahren,* Arch. Rational Mech. Anal. **6** (1960), 261–276. MR **27** #3069.

169. ———, *Fehlerabschätzungen mit Hilfe eines Vergleichssatzes,* Z. Angew. Math. Mech. **42** (1962), 187–193. MR **25** #756.

170. ———, *Ein Vergleichssatz unter Verwendung höherer Ableitungen,* Z. Angew. Math. Mech. **43** (1963), 81–83. MR **26** #5425.

171. ———, *Zur Fehlerabschätzung näherungsweiser Losungen von Gleichungen in halbgeordneten Räumen,* Arch. Math. **14** (1963), 130–138. MR **26** #7141.

172. J. Schröder, *Das Iterationsverfahren bei allgemeinerem Abstandsbegriff,* Math. Z. **66** (1956), 111–116. MR **18**, 765.

173. ———, *Neue Fehlerabschätzungen für verschiedene Iterationsverfahren,* Z. Angew. Math. Mech. **36** (1956), 168–181. MR **18**, 152.

174. ———, *Nichtlineare Majoranten beim Verfahren der schrittweisen Näherung,* Arch. Math. **7** (1957), 471–484. MR **19**, 460.

175. S. V. Simeonov, *A successive approximation process and its application in solving functional equations with non-linear operators of monotonic type,* Dokl. Akad. Nauk SSSR **138** (1961), 1033–1034 = Soviet Math. Dokl. **2** (1961), 790–791. MR **25** #3390.

176. ———, *On a process of successive approximations and its application to the solution of functional equations with nonlinear operators of monotone type,* Godisnik Inž.-Stroit. Inst. Fak. Stroit. Arhitekt. i Hidrotehn. **14** (1962), no. 1, 9–21 (1963). (Bulgarian) RŽMat. **1965** #7Б465.

177. ———, *The application of a process of successive approximations to the solution of certain types of functional equations,* Dokl. Akad. Nauk SSSR **148** (1963), 534–537 = Soviet Math. Dokl. **4** (1963), 144–147. MR **27** #489.

178. V. H. Sirenko, *On the numerical realization of the method of averaging functional corrections,* Ukrain. Mat. Ž. **13** (1961), no. 4, 51–66. (Russian) MR **26** #2030.

179. S. N. Slugin, *Approximate solution of operator equations on the basis of S. A. Čaplygin's method,* Dokl. Akad. Nauk SSSR **103** (1955), 565–568, 746. (Russian) MR **17**, 387.

180. N. S. Smirnov, *Introduction to the theory of nonlinear integral equations*, ONTI, Moscow, 1936. (Russian)

181. V. I. Smirnov, *A course in higher mathematics*. Vol. 5, rev. ed., Fizmatgiz, Moscow, 1959; English transl., Pergamon Press, Oxford; Addison-Wesley, Reading, Mass., 1964. MR 25 #5141; 29 #5964.

182. S. L. Sobolev, *Applications of functional analysis in mathematical physics*, Izdat. Leningrad. Gos. Univ., Leningrad, 1950; English transl., Transl. Math. Monographs, vol. 7, Amer. Math. Soc., Providence, R.I., 1963. MR 14, 565; 29 #2624.

183. Ju. D. Sokolov, *On a problem of the theory of unsteady motion of ground water*, Ukrain. Mat. Ž. 5 (1953), 159–170. (Russian) MR 15, 476.

184. ———, *On the theory of plane unsteady filtration of ground water*, Ukrain. Mat. Ž. 6 (1954), 218–232. (Russian) MR 18, 91.

185. ———, *On approximate solution of the basic equation of the dynamics of a hoisting cable*, Dopovidī Akad. Nauk Ukraïn. RSR 1955, 21–25. (Ukrainian) MR 17, 307.

186. ———, *On an axially symmetric problem of the theory of unsteady motion of ground water*, Ukrain. Mat. Ž. 7 (1955), 101–111. (Russian) MR 18, 91.

187. ———, *On a method of approximate solution of linear integral and differential equations*, Dopovidī Akad. Nauk Ukrain. RSR 1955, 107–111. (Ukrainian) MR 17, 196.

188. ———, *On the determination of dynamic pull in shaft-lifting cables*, Akad. Nauk Ukrain. RSR Prikl. Meh. 1 (1955), 23–25. (Ukrainian) MR 18, 609.

189. ———, *Sur la méthode du moyennage des corrections fonctionnelles*, Ukrain. Mat. Ž. 9 (1957), 82–100. (Russian) MR 19, 687.

190. ———, *Sur l'application de la méthode des corrections fonctionnelles moyennes aux équations intégrales non linéaires*, Ukrain. Mat. Ž. 9 (1957), 394–412. (Russian) MR 21 #2888.

191. ———, *Sur la résolution approchée des équations intégrales linéaires du type de Volterra*, Ukrain. Mat. Ž. 10 (1958), no. 2, 193–208. (Russian) MR 20 #7393.

192. ———, *Sur une méthod de la résolution approchée des équations intégrales non linéaires à limites variables*, Ukrain. Mat. Ž. 10 (1958), 419–433. (Russian) MR 21 #977.

193. ———, *Sur l'application de la méthode des corrections fonctionnelles moyennes aux équations du type parabolique linéaries par rapport aux dérivées*, Ukrain. Mat. Ž. 12 (1960), 181–195. (Russian) MR 28 #4315.

194. ———, *On a method of approximate solution of systems of linear integral equations*, Ukrain. Mat. Ž. 13 (1961), no. 4, 79–87. (Russian) MR 26 #2031.

195. ———, *On a method of approximate solution of systems of nonlinear integral equations with constant limits*, Ukrain. Mat. Ž. 15 (1963), 58–70. (Russian) MR 26 #5385.

196. ———, *On sufficient tests for the convergence of the method of averaging functional corrections*, Ukrain. Mat. Ž. 17 (1965), no. 3, 91–103. (Russian) MR 33 #1770;

197. ———, *The method of averaging of functional corrections*, "Naukova Dumka", Kiev, 1967. (Russian) MR 35 #3931.

198. A. A. Stonickiĭ, *On the solution of some classes of linear differential and integro-differential equations with small parameter*, Candidate's Dissertation, Inst. Mat. Akad. Nauk Ukrain. SSR, Kiev., 1963. (Russian)

199. A. A. Stonickiĭ, *Approximate solution, by the method of Ju. D. Sokolov, of an infinite system of integral equations of Volterra type depending on a parameter*, Dopovīdī Akad. Nauk Ukraïn. RSR 1963, 1555–1559. (Ukrainian)   MR 29 #2610.

200. I. T. Švec', V. Ī. Fedorov and V. G. Bodnarčuk, *Application of some approximate methods to the solution of the equation of heat conduction in the rotors of a turbine*, Zb. Prac' Inst. Teploenerget. Akad. Nauk Ukrain. SSR Vyp. 18 (1960), 3–15. (Ukrainian)

201. A. Tarski, *A lattice-theoretical fixpoint theorem and its applications*, Pacific J. Math. 5 (1955), 285–309.   MR 17, 574.

202. A. N. Tihonov, *Ein Fixpunktsatz*, Math. Ann. 111 (1935), 767–776.

203. V. I. Tivončuk, *On the application of the method of averaging functional corrections to the solution of linear integral equations of Volterra type*, Proc. Sci. Conf. Engrs., Aspirants and Junior Research Assistants Inst. Math. Acad. Sci. Ukrain. SSR, Izdat. Inst. Mat. Akad. Nauk Ukrain. SSR, Kiev, 1963, pp. 73–76.   (Russian)

204. ———, *Application of the method of Ju. D. Sokolov to the solution of linear integral equations of mixed type*, Dopovīdī Akad. Nauk Ukrain. RSR 1964, 1014–1018. (Ukrainian)   MR 30 #2304.

205. ———, *An error bound for a variant of Ju. D. Sokolov's method of solving linear integral equations of Volterra type and equations of mixed type*, Dopovīdī Akad. Nauk Ukraïn. RSR 1964, 1281–1284. (Ukrainian)   MR 30 #2305.

206. ———, *On the solution of linear integral equations of mixed type by means of a variant of the method of Ju. D. Sokolov*, Dopovīdī Akad. Nauk Ukrain. RSR 1964, 1559–1563. (Ukrainian)   MR 30 #3347.

207. ———, *The solution of linear integral equations of Volterra type by a variant of the method of Ju. D. Sokolov*, Ukrain. Mat. Ž. 17 (1965), no. 1, 77–88. (Russian)   MR 32 #8083.

208. ———, *On the solution of Volterra linear integral equations and equations of mixed type in L^p-space by a variant of the method of Ju. D. Sokolov*, Ukrain. Mat. Ž. 17 (1965), no. 4, 133–139. (Russian)   MR 33 #3062.

209. ———, *On the application of the method of averaging functional corrections to the solution of linear Volterra integral equations and linear integral equations of mixed type*, Candidate's Dissertation Inst. Mat. Akad. Nauk Ukrain. SSR, Kiev, 1965. (Russian)

210. ———, *Solution of linear integral equations of Volterra type and equations of mixed type by means of a variant of a method of Ju. D. Sokolov*, First Republ. Math. Conf. of Young Researchers, part I, Akad. Nauk Ukrain. SSR Inst. Mat., Kiev, 1965, pp. 628–635. (Russian)   MR 33 #6317.

211. ———, *A variant of the method of averaging functional corrections for solution of linear integral equations of mixed type*, Differencial'nye Uravnenija 2 (1966), 1228–1238 = Differential Equations 2 (1966), 636–641.   MR 34 #4841.

212. ———, *On a variant of the method of averaging functional corrections for the solution of a linear integral equation of mixed type*, Proc. Second Conf. Young Ukrainian Mathematicians, "Naukova Dumka", Kiev, 1966, pp. 601–605. (Ukrainian)

213. J. Todd (Editor), *Survey of numerical analysis*, McGraw-Hill, New York, 1962. MR 24 #B1271.

214. Marko Todorow, *Über die iterative Behandlung linearer Gleichungssysteme*, Bautechnik 35 (1958), no. 4, 136–138. RŽMat. 1959 #877.

215. F. G. Tricomi, *Integral equations,* Pure and Appl. Math., vol. 5, Interscience, New York, 1957. MR 20 #1177.

216. N. I. Tukalevskaja, *Numerical realization of the method of averaging functional corrections for integral equations of Volterra type,* Proc. Sci. Conf. Engrs., Aspirants and Junior Research Assistants Inst. Math. Acad. Sci. Ukrain. SSR, Izdat. Inst. Mat. Akad. Nauk Ukrain. SSR, Kiev, 1963, pp. 77–83. (Russian)

217. ———, *On a certain method of solving linear integral equations of Volterra type,* Dopovīdī Akad. Nauk Ukraïn. RSR 1965, 998–1002. (Ukrainian) MR 33 #5155.

218. ———, *Approximate solution of linear integral equations of Volterra type,* First Republ. Math. Conf. of Young Researchers, part I, Akad. Nauk Ukrain. SSR Inst. Mat., Kiev, 1965, pp. 636–640. (Russian) MR 33 #6318.

219. ———, *On a certain method of approximate solution of Volterra type linear integral equations in the class of L^p-functions,* Dopovīdī Akad. Nauk Ukraïn. RSR 1966, 299–302. (Ukrainian) MR 33 #4629.

220. ———, *On a method of approximate solution of linear Volterra integral equations and equations of mixed type,* Candidate's Dissertation, Inst. Mat. Akad. Nauk Ukrain. SSR, Kiev, 1966. (Russian)

221. ———, *On a method of approximate solution of linear integral equations of mixed type in L^p-spaces,* Proc. Second Sci. Conf. Young Ukrainian Mathematicians, "Naukova Dumka", Kiev, 1966, p. 609. (Ukrainian)

222. N. I. Tukalevskaja and A. V. Nesterčuk, *On a method of solving linear integral equations of Volterra type,* Ukrain. Mat. Ž. 17 (1965), no. 1, 95–101. (Russian) MR 32 #8084.

223. M. M. Vaĭnberg, *Variational methods for the study of non-linear operators,* GITTL, Moscow, 1956; English transl., Holden-Day, San Francisco, Calif., 1964. MR 19, 567; 31 #638.

224. ———, *On the convergence of the method of steepest descents for non-linear equations,* Dokl. Akad. Nauk SSSR 130 (1960), 9–12 = Soviet Math. Dokl. 1 (1960), 1–4. MR 25 #571.

225. G. M. Vaĭnikko, *On the question of convergence of Galerkin's method,* Tartu Riikl. Ül. Toimetised Vih. 177 (1965), 148–153. (Russian) MR 36 #1094.

226. ———, *The perturbed Galerkin method and the general theory of approximate methods for nonlinear equations,* Ž. Vyčisl. Mat. i Mat. Fiz. 7 (1967), 723–751 = USSR Comput. Math. and Math. Phys. 7 (1967), no. 4, 1–41. MR 36 #1095.

227. Vo-Khac Khoan, *Q-solutions d'un système différentiel,* C. R. Acad. Sci. Paris 258 (1964), 3430–3433. MR 29 #500.

228. Ju. V. Vorob'ev, *Method of moments in applied mathematics,* Fizmatgiz, Moscow, 1958; English transl., Gordon and Breach, New York, 1965. MR 21 #7591; 32 #1872.

229. B. Z. Vulih, *Introduction to theory of partially ordered spaces,* Fizmatgiz, Moscow, 1961; English transl., Noordhoff, Groningen, 1967. MR 24 #A3494; 37 #121.

230. Sheng-wang Wang, *On the solutions of the Hammerstein integral equations,* Bull. Acad. Polon. Sci. Sér. Sci. Math. Astronom. Phys. 8 (1960), 339–342. MR 23 #A2727.

231. J. Warga, *On a class of iterative procedures for solving normal systems of ordinary differential equations,* J. Math. Physics 31 (1953), 223–243. MR 14, 587.

232. T. Ważewski, *Sur la méthode des approximations successives*, Ann. Soc. Polon. Math. 16 (1938), 214–215.

233. ———, *Sur une extension du procédé de I. Jungermann pour établir la convergence des approximations successives au cas des équations différentielles ordinaires*, Bull. Acad. Polon. Sci. Sér. Sci. Math. Astronom. Phys. 8 (1960), 43–46. MR 23 #A3404.

234. ———, *Sur un procédé de prouver la convergence des approximations successives sans utilisation des séries de comparison*, Bull. Acad. Polon. Sci. Sér. Sci. Math. Astronom. Phys. 8 (1960), 47–52. MR 23 #A3405.

235. J. Weissinger, *Zur Theorie und Anwendung des Iterationsverfahrens*, Math. Nachr. 8 (1952), 193–212. MR 14, 478.

236. R. Zuber, *A certain algorithm for solving ordinary differential equations of the first order.* I, Zastos. Mat. 8 (1966), 351–363. (Polish) MR 33 #6844.

237. ———, *A certain algorithm for solving ordinary differential equations of the first order.* II, Zastos. Mat. 9 (1966), 85–98. (Polish) MR 34 #3799.

238. ———, *A method of successive approximation*, Bull. Acad. Polon. Sci. Sér. Sci. Math. Astronom. Phys. 14 (1966), 559–561. MR 35 #5684.